Robert Shufflebotham

InDesign

Third Edition

For Windows and Mac

In easy steps is an imprint of In Easy Steps Limited
16 Hamilton Terrace · Holly Walk · Leamington Spa
Warwickshire · United Kingdom · CV32 4LY
www.ineasysteps.com

Third Edition

Notice of Liability
Every effort has been made to ensure that this book contains accurate
and current information. However, In Easy Steps Limited and the
author shall not be liable for any loss or damage suffered by readers
as a result of any information contained herein.

Trademarks
Adobe® and InDesign® are registered trademarks of Adobe Systems
Incorporated. All trademarks are acknowledged as belonging to their
respective companies.

In Easy Steps Limited supports The Forest Stewardship Council (FSC),
the leading international forest certification organization. All our titles
that are printed on Greenpeace approved FSC certified paper carry the
FSC logo.

MIX
Paper from
responsible sources
FSC® C020837

Printed and bound in the United Kingdom

ISBN 978-1-84078-936-2

Contents

1 The Working Environment

This chapter takes a look at
the working environment
and gets you started using
Adobe InDesign. It covers
setting up a new document
and introduces tool and
panel conventions, along
with other useful techniques
that will help make you
accurate and productive as
you start to use the software.

The Home Screen

Click the **Learn** button to show a list of web-based tutorials from Adobe.

Click the **Create new** button to go to the New Document dialog box. Click the **Open** button to open an existing InDesign document using the standard Windows/Mac Open dialog box.

Use the **Sort** pop-up menu to control the ordering of the thumbnails. **Recent** is the default: your most recently opened documents appear at the start of the list. You can also sort by Name, Size and Kind.

The **Sort Order button** (▼) is a toggle to reverse the order in which thumbnails appear – for example, from largest file size to smallest; or smallest to largest.

Thumbnails of the most recent InDesign documents you've worked on appear in the **Recent** pane, giving you quick visual access to projects on which you are currently working.

Click on a thumbnail to open the InDesign document.

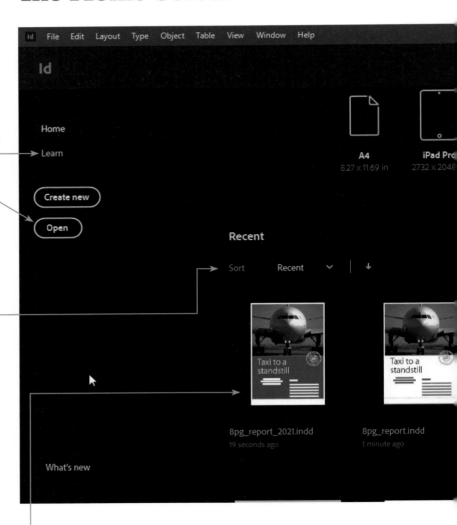

The author

Robert Shufflebotham draws on over 28 years' experience of working in top-flight publishing, design and marketing environments. He is a qualified teacher.

He has trained and written books on InDesign since version 1. During this time he has developed a unique understanding of what it takes to get the best out of InDesign from the start.

Benefit from his considerable and varied experience distilled in the pages of this book as he guides you through the essential functionality that puts you on course to successfully master the software.

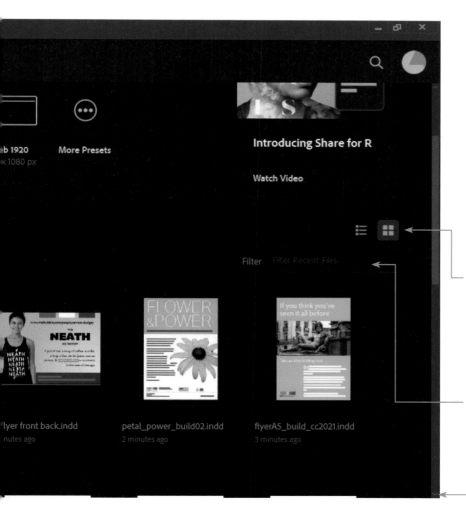

Click the **Hide suggestions** button (in the top-right corner of the Home screen) to hide the suggested preset document sizes – A4/Letter, iPad Pro, Web – and the New Features videos pane:

^ Hide suggestions

Hiding suggestions can free up more space on your screen to show more InDesign document thumbnails.

Use the **List/Grid** buttons to control the appearance of the thumbnails. Grid view provides additional information such as file size and kind.

Use the **Filter** entry box to quickly locate InDesign documents by name.

Scroll down to see all available thumbnails.

Learning InDesign

This book offers a unique approach to learning InDesign. Its prime purpose is to focus on the fundamental principles and core processes, common to all versions, that you need to know as you start using InDesign in order to become a confident, proficient and productive user. Master these fundamentals and you can progress to exploit the full creative potential of a software application that is undoubtedly one of the most powerful, creative tools available.

Because of this clear focus and an approach

that breaks down the software into easily digested learning sequences, you can use this book to comfortably learn to use all recent versions of the software.

Importantly, **InDesign in easy steps** does not seek to cover every single function in InDesign. Such attempts can lead to information overload for a new user – all too often it's not quite clear what is essential and what is of marginal use, and as a result, the new user cannot develop a clear, logical learning strategy for understanding and then mastering the new software.

New Document Dialog Box

Selecting an Intent

When you create a new document, one of the first choices you need to make is to select your Intent.

Select an Intent according to your primary final output requirements: is your InDesign document going to be printed or is it going to be used online or displayed on a smartphone?

If you select **Print**, the default color mode is set to **CMYK** (see pages 106-110 for information on color modes). The measurement system will typically be inches in the US, or millimeters/centimeters in Europe and the UK. The Blank Document Presets are standard print-size templates designed for printing.

Select **Web** or **Mobile** to see standard Blank Document screen sizes in pixels, with RGB set as the default color mode. Notice in the Preset Details panel on the right that the **Units** setting changes automatically to **Pixels** if you select **Web** or **Mobile**.

Document Preset thumbnails

New Document

Recent Saved **Print** Web Mobile

BLANK DOCUMENT PRESETS (10)

A4
210 x 297 mm

Letter
215.9 x 279.4 mm

Legal
215.9 x 355.6 mm

View All Presets +

TEMPLATES (71)

Art and Culture Mag... FREE

Basic Style Guide La... FREE

Black and White Eve...

Find more templates on Adobe Stock

Template thumbnails

Click **View All Presets** + to see an
expanded set of available presets.

PRESET DETAILS

Untitled-2

Width
8.5 in

Units
Inches

Height
11 in

Orientation

Pages
1

Facing Pages
☑

Start #
1

Primary Text Frame
☐

Columns
1

Column Gutter
0.1667 in

∨ Margins

Top
0.5 in

Bottom
0.5 in

Inside
0.5 in

Outside
0.5 in

Tabloid
279.4 x 431.8 mm

Bold Resume and Co... FREE

Go

☐ Preview   **Create**   Close

Document setup options
can be customized in the
PRESET DETAILS pane (see
pages 12-13 for details).

The **Ctrl** key (Windows)
and the **Command key**
(sometimes referred to as
the "Apple" key on Mac
keyboards),
and the **Alt** key (Windows)
and the **Option** ⌥ (Mac)
key are used identically as
modifier keys.

The **Shift** key is standard
on both platforms.

The **Control** key is specific
to Macs.

When you are
already working
in an InDesign
document,
choose

File > New > Document
(**Ctrl/Command + N**) to
access the New Document
dialog box.

Depending on the size of your
computer screen, you may need
to scroll down to see all the
document setup options.

Document Setup

When you click on a Blank Document preset, the settings in the Preset Details panel update according to the preset you selected. You can edit all the settings in the Preset Details panel to create the exact page settings for your design.

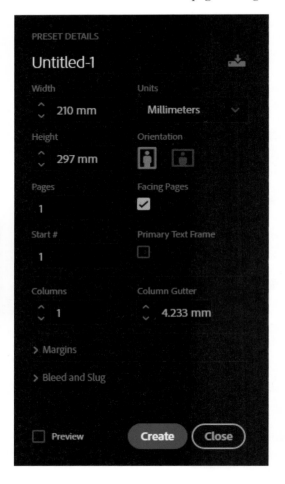

Size and Orientation

1. To create a custom page size, enter the exact dimensions you want in the **Width** and **Height** fields.

2. Use the **Units** pop-up menu to select the measurement system you want to use.

3. Click the **Landscape** icon to create a landscape page orientation. Portrait is selected by default.

4. Facing Pages is selected by default. Use the **Facing Pages** option to create a publication such as a magazine, brochure or book that will consist of double-page spreads. With **Facing Pages** selected, Left and Right in the Margins area change to Inside and Outside, allowing you to set a wider inside margin to accommodate any binding edge in your document.

5. Enter the number of pages you want in the document in the **Pages** entry box. If you are unsure of the number of pages you need, you can always add or delete pages when you are working in the document.

6. Use the **Start #** entry field to start a document with a left-hand page (in which case you enter an even number), or if you want to create a section as you set up a document.

Bleed and Slug

A **bleed** is required in printing when you want an area of color, or an image, to run right to the edge of the final trimmed page size. The bleed essentially allows for a margin of error during the printing and trimming process. Ask your commercial printer for their required bleed amount before you start your document.

Enter values for **Slug** to define an area outside the page area and bleed where instructions to the printer or other information relating to the document, such as title and date, can appear.

Both the slug and bleed areas are trimmed and do not appear on the final printed page.

For example, if you enter an even number, InDesign creates a left-hand page to begin the document, instead of the default right-hand page. If you subsequently set up automatic page numbering (see pages 130-131 for further details), the pages start numbering from the value you enter in this field.

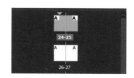

7 Select the **Primary Text Frame** option to create an automatic text frame on the A-master that fits to the margins and matches the number columns and gutter specified in the dialog box.

The **Primary Text Frame** option is useful if you want to create a publication with a regular, rigid text frame requirement – for example, a novel or a training manual.

Margin Guides

Set margin guides to define the main text area of your document. This is often referred to as the "safe text area". Margin guides are non-printing and appear as magenta lines on the screen. Margin guides are only guides – objects can be placed across margin guides or completely outside the margin guides.

1 Enter values for **Top**, **Bottom**, **Left/Right**, **Inside/Outside**.

2 Alternatively, click the **Make all settings the same** button () to make it active, and then enter a value in one of the **Margin** entry fields. Press the **Tab** key to make all values the same as the first value you enter.

Column Guides

Column guides are non-printing and appear on screen as purple lines. They serve as a grid for constructing a publication. You are not constrained to working within the columns – all InDesign objects can cross column guides as necessary to create the page design you require.

1 Enter the number of columns you want. Specify a **Column Gutter** – the space between columns. InDesign calculates the width of columns based on the overall width of the page, the number of columns, and the values of margins and gutters.

2 When working in a document, to change margin and column settings for the currently selected master page, spread or page, choose **Layout > Margins and Columns**. Enter new values as required. **OK** the dialog box.

The Working Environment

The Adobe InDesign interface is virtually identical on both the Windows and Macintosh platforms, making it easy to work in both environments without the need for extensive retraining.

This book uses a mixture of Windows and Macintosh screenshots, and the instructions given apply equally to both platforms.

Home button

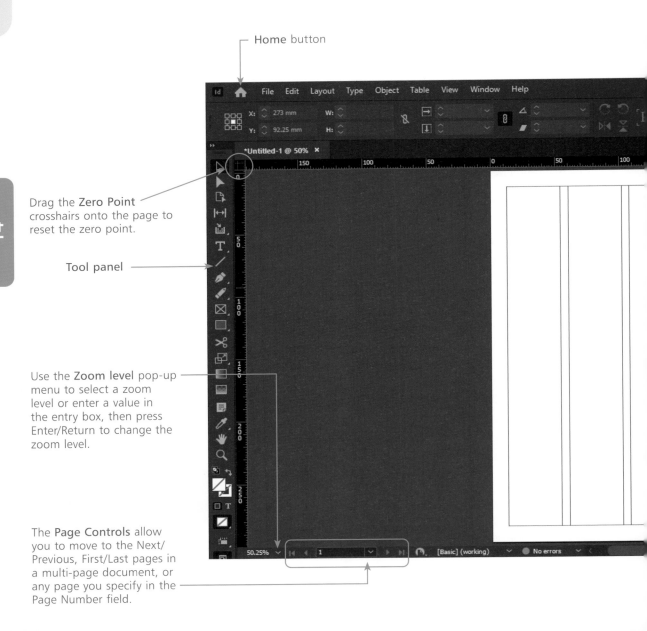

Drag the **Zero Point** crosshairs onto the page to reset the zero point.

Tool panel

Use the **Zoom level** pop-up menu to select a zoom level or enter a value in the entry box, then press Enter/Return to change the zoom level.

The **Page Controls** allow you to move to the Next/ Previous, First/Last pages in a multi-page document, or any page you specify in the Page Number field.

Based on over 20 years' experience teaching and using Adobe InDesign, the **Essentials Classic** workspace provides the best balance between ease of use and access to features for new users whose aim is to develop their design and publishing skills to produce high-quality, creative results.

As you gain confidence and develop your InDesign skillset, you should consider moving to the **Advanced** workspace.

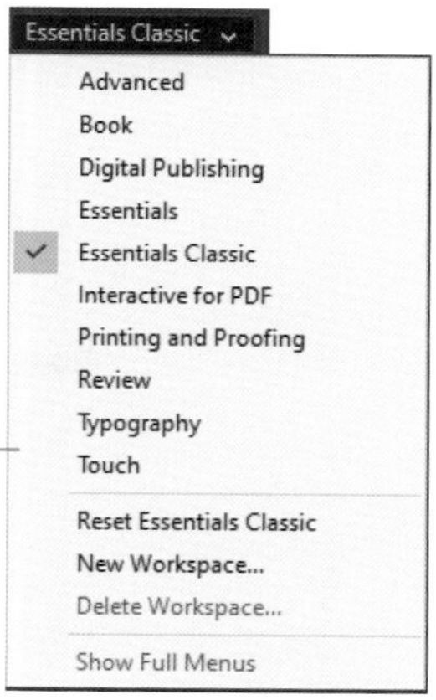

Essentials Classic ∨
Advanced
Book
Digital Publishing
Essentials
✓ Essentials Classic
Interactive for PDF
Printing and Proofing
Review
Typography
Touch
Reset Essentials Classic
New Workspace...
Delete Workspace...
Show Full Menus

The **Control panel** (not visible in the Essentials workspace) provides immediate access to a huge variety of powerful, essential InDesign controls.

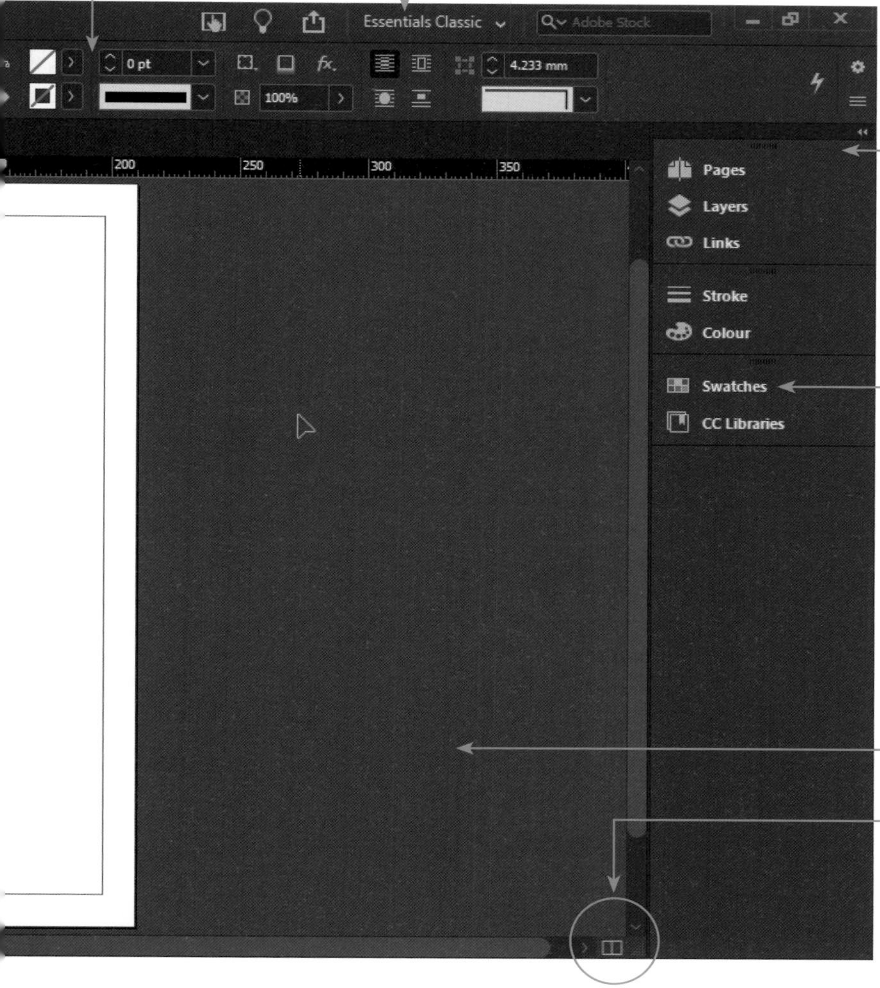

Workspace pop-up

Panel Dock – Essentials Classic workspace

Click on a panel icon in the Panel Dock to expand the panel, along with other panels in the same panel group.

Pages
Layers
Links
Stroke
Colour
Swatches
CC Libraries

The **Pasteboard** area exists all around the document page or spread. Objects you place on the Pasteboard are saved with the document but do not print. It is a good practice to clear the Pasteboard of all unused objects prior to final output to avoid any possibility of confusion.

Click the **Scroll arrows** to scroll the page up, down, left or right in increments. You can drag the **Scroll Box** to move the page a custom amount. Alternatively, click either side of the Scroll Box to move the window in half-screen increments.

The Tool Panel

Use the following techniques to choose tools and to work quickly and efficiently as you build page layouts in Adobe InDesign.

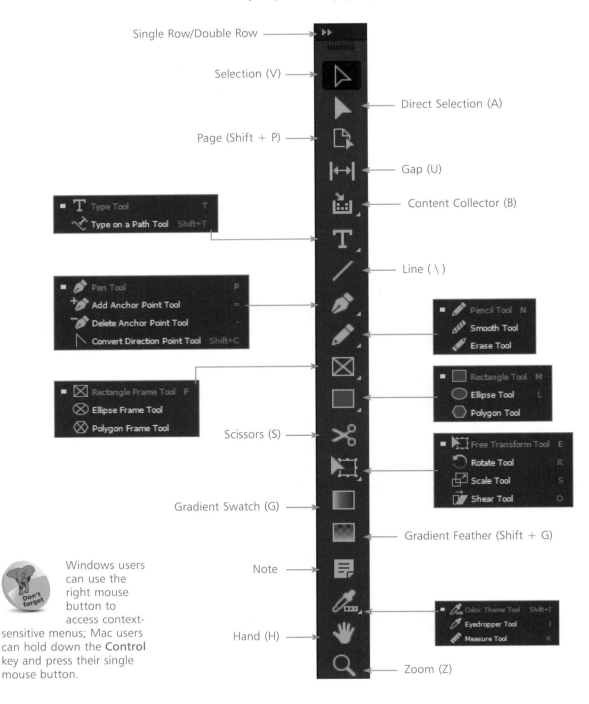

Single Row/Double Row

Selection (V)

Direct Selection (A)

Page (Shift + P)

Gap (U)

Content Collector (B)

Type Tool — T
Type on a Path Tool — Shift+T

Line (\ )

Pen Tool — P
Add Anchor Point Tool — =
Delete Anchor Point Tool — −
Convert Direction Point Tool — Shift+C

Pencil Tool — N
Smooth Tool
Erase Tool

Rectangle Frame Tool — F
Ellipse Frame Tool
Polygon Frame Tool

Rectangle Tool — M
Ellipse Tool — L
Polygon Tool

Scissors (S)

Free Transform Tool — E
Rotate Tool — R
Scale Tool — S
Shear Tool — O

Gradient Swatch (G)

Gradient Feather (Shift + G)

Note

Color Theme Tool — Shift+I
Eyedropper Tool — I
Measure Tool — K

Hand (H)

Zoom (Z)

Don't forget

Windows users can use the right mouse button to access context-sensitive menus; Mac users can hold down the **Control** key and press their single mouse button.

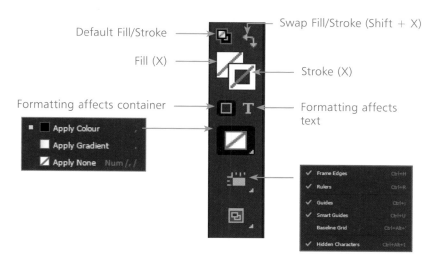

Default Fill/Stroke

Swap Fill/Stroke (Shift + X)

Fill (X)

Stroke (X)

Formatting affects container

Formatting affects text

- Apply Colour
- Apply Gradient
- Apply None Num /. /

✓ Frame Edges   Ctrl+H
✓ Rulers   Ctrl+R
✓ Guides   Ctrl+;
✓ Smart Guides   Ctrl+U
  Baseline Grid   Ctrl+Alt+'
✓ Hidden Characters Ctrl+Alt+I

Tool Panel Techniques

1 Most of the time you will have the Tool panel visible as you build documents. If you accidentally close it, choose **Window > Tools** to display the Tool panel.

2 To choose a tool, click on it in the Tool panel. The tool is highlighted and when you move your cursor back into the InDesign window, the cursor changes to indicate the tool you selected.

3 A small, gray triangle in the bottom-right corner of a tool icon indicates that there are additional tools available in the tool group. To access a hidden tool, press and hold the tool currently showing in the Tool panel; this will show the tool group pop-up. Slide your cursor onto the tool you want to select, and then release. The tool you select is displayed in the Tool panel as the default tool in that group until you choose another tool from the group.

4 Provided that you do not have the Text Insertion Point located in the text, press the **Tab** key to hide all visible panels, including the Tool panel. Press **Tab** again to show all previously visible panels. Hold down **Shift** and press the **Tab** key to hide all panels except the Tool panel.

5 Rest your cursor on a tool in the Tool panel for a few seconds to display the Tool Tip label. This tells you the name of the tool and, in brackets, its keyboard shortcut.

Hot tip

Double-click the **Eyedropper** tool to access the Eyedropper Options dialog box, where you can create settings to control exactly which attributes the Eyedropper copies:

Eyedropper Options

> ✓ Stroke Settings
> ✓ Fill Settings
> ✓ Character Settings
> ✓ Paragraph Settings
> ✓ Object Settings
> ☐ Transform Options

17

Hot tip

If you find the Tool Tip labels distracting, you can turn them off: choose **Edit > Preferences > Interface** (Windows) or **InDesign > Preferences > Interface** (Mac), then change the Tool Tips pop-up to **None** in the **Cursor and Gesture Options** section.

Opening Documents

As well as creating new InDesign documents, you will often need to open existing documents. If you open an InDesign document created in an earlier version of the software, InDesign converts the document to the version you are using and adds "[Converted]" to the document tab. When you save a converted file you are prompted to perform a Save As operation to save an updated copy of the file using the newer version of the software.

InDesign documents have the extension .indd. InDesign templates have the extension .indt.

1 To open an existing document, choose **File** > **Open**.

2 Use standard Windows/Mac techniques to navigate to the file you want to open.

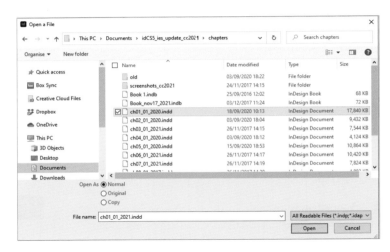

3 Click the file name to select it. Click the **Open** button.

4 In the **Open a File** dialog box, click the **Copy** radio button to open a copy of the file – the file opens as an untitled document.

5 When you want to edit an InDesign template, in the **Open a File** dialog box, click on the template file name and then select the **Original** radio button before you click **Open**.

Choose **File** > **Open Recent** to access a list of your most recently used InDesign files.

 When you are working on an InDesign document, you can click the **Home** button to return to the Home screen (see pages 8-9).

 From the Home screen, click the **Back to InDesign** button to return to any active InDesign document you have open.

Zooming and Scrolling

Use the **Zoom tool** in the **Zoom Level** pop-up menu and a variety of keyboard shortcuts for changing the zoom level as you work on a document. You can use the **Hand tool** and the scroll bars for moving around the document.

1 To use the **Zoom tool**, select it, position the Zoom cursor in the document window, and then click to zoom in at the cursor position by preset increments. With the **Zoom tool** selected, hold down **Alt/Option** to change the cursor to the zoom out cursor. Click to zoom out by preset decrements.

2 One powerful zoom technique is to press and drag with the **Zoom tool** to define an area on which you want to zoom. A dotted rectangle appears, defining the zoom area. The smaller the zoom area you define, the greater the resulting magnification.

you can't beat us

3 You can use standard scroll bar techniques to see different parts of a page, or you can use the **Hand tool**: select the **Hand tool**, position your cursor on the page, and then press and drag to scroll the page.

4 You can zoom to preset percentage levels using the **Zoom Level** pop-up menu in the Application bar along the top edge of the InDesign workspace.

5 The **View** menu offers standard options for changing the magnification of the page. The keyboard shortcut is listed with each option.

5%	
12.5%	
25%	
50%	
75%	
100%	
125%	
150%	
200%	
300%	
400%	
600%	
800%	
1200%	
1600%	
2400%	
3200%	
4000%	
65.25%	

Zoom In	Ctrl+=
Zoom Out	Ctrl+-
Fit Page in Window	Ctrl+0
Fit Spread in Window	Ctrl+Alt+0
Actual Size	Ctrl+1
Entire Pasteboard	Ctrl+Alt+Shift+0

With any tool other than the **Zoom tool** selected, hold down **Ctrl/ Command** + **Spacebar** to temporarily access the **Zoom tool**. Add **Alt/ Option** to the above combination to access the zoom-out option.

With any tool other than the **Hand tool** selected, hold down the **Spacebar** on the keyboard to gain temporary access to the **Hand tool**. If you are working with the **Type tool**, hold down **Alt/Option** to access the **Hand tool**, otherwise you will introduce spaces at the Text Insertion Point.

Workspaces and Panels

Don't forget

Your workspace is the arrangement of panels and document windows visible as you work in InDesign. There is a wide variety of techniques for managing your workspace so that you can work comfortably, efficiently and in a way that suits your own individual preferences.

Hot tip

You can save a workspace so that you can quickly and easily recreate a particular arrangement of panels. To do this, go to the Workspace pop-up menu and select **New Workspace**. Enter a name for your custom workspace, then click **OK**. The custom workspace appears in the list of workspaces.

Hot tip

To move a floating panel into the Panel Dock, position your cursor in the gray bar at the top of the panel, then drag the panel into the Dock. A bright blue bar indicates where the panel will appear when you release the mouse button.

Workspaces

Most of InDesign's functionality is accessed through panels that contain groups of related controls. Panels give you quick and easy access to the controls you need to use most frequently. Panels are grouped together in a "Panel Dock" running down the right-hand side of the InDesign window.

The panels you see initially are determined by the active workspace. The default workspace is the Essentials workspace, but you can select a different workspace, offering a different set of panels, to suit your needs.

1 Use the **Workspace** pop-up menu to choose a default arrangement of panels to suit your requirements. You can modify a default workspace using the techniques outlined below to create a custom workspace.

2 To revert a modified workspace to the default arrangement, you can select **Reset [name of workspace]** from the **Workspace** pop-up menu.

3 Additional panels, not displayed in the Panel Dock, are available from the Window menu – for example, choose **Window > Info** to show the Info panel. When you show such a panel from the Window menu, it appears as a "floating panel". You can move floating panels around on your screen, or you can attach them in the Panel Dock.

Managing Panels and the Panel Dock

The following techniques will help you customize the way you work with panels and the Panel Dock.

1 Click the **Expand panels** button () to display expanded panel groups. Click the **Collapse to Icons** button ( ) to shrink the dock to icons and labels. (This screenshot shows the Advanced workspace Panel Dock.)

Position your cursor on the left edge of the Panel Dock. When the cursor changes to the bidirectional arrow (⇔), drag to the right to create an icons-only dock. Drag to the left to expand the dock manually.

2 For the Panel Dock in workspaces showing icon labels, click the panel name to display the panel as a pop-out panel.

3 For an expanded dock, to make a panel active, click its tab. The tab highlights, and the options for the panel are displayed.

The Essentials Workspace Panel Dock

The Essentials workspace Panel Dock provides three main tabs along the top: Properties, Pages and CC Libraries. The Properties tab contains a variety of context-related panes, where the panes that are available change depending on what you have selected in your InDesign document:

Panels, other than the Properties panel in the Essentials workspace, have a **panel menu** button (☰) allowing you to access additional commands:

Some panes in the Properties panel – for example, Transform and Text Wrap – have a **More Options** button (⋯) to display an extended set of related controls.

Properties panel –
nothing selected

Properties panel –
Graphic Frame selected

Properties panel –
Text Frame selected

The Control Panel

The **Control panel** is available as a default panel for the Advanced, Essentials Classic, Book, Digital Publishing, Interactive for PDF and Typography workspaces. (See pages 20-21 for information on selecting a workspace.)

The Control panel is one of the most versatile panels in InDesign. The default position for the panel is docked below the Menu bar. It provides convenient access to an extensive range of settings and controls that change depending on the tool you are working with and the object you have selected.

1 When you are working in a text frame with the **Type tool** selected, you can choose between the **Character Formatting Controls** button and the **Paragraph Formatting Controls** button as required.

Graphic/Shape frame selected with Selection tool

Text frame selected with Type tool

Panels in workspaces other than Essentials have a **panel menu** button (■), giving you access to additional commands.

"Quick set" document preferences

When working with the Control panel, or any other panel or dialog box in InDesign, if you make a change to a setting **with nothing selected**, that setting becomes the default for objects you subsequently create.

For example, if you use options in the Control panel to change the font to Arial and the type size to 72pt, the next time you create a text frame, text you enter appears with these settings.

This is a useful and powerful feature, but in the early stages of learning InDesign, be careful that it doesn't catch you out.

Remember – these preferences are specific to the document you are working in.

If you are setting up templates for professional publications such as books, magazines and newspapers, be very careful as to which paragraph, character and object styles are set when you create a template.

Ruler Guides

Ruler guides are non-printing guides that are used to align objects accurately. The default color for ruler guides is light blue. When **Snap to Guides** is **On**, Drawing tool cursors snap onto guides when they come within four screen pixels of the guide. Also, when you move an object, the edges of the object will snap onto guides.

1. To create a ruler guide, make sure the rulers are showing; choose **View > Show Rulers (Ctrl/Command + R)** if they are not. Position your cursor in either the top or the left ruler then press and drag onto the page. When you release, you create a ruler guide at that point. Look

   at the Control panel or the Transform panel X/Y fields to get a readout indicating the position of the cursor as you press and drag.

2. To create a ruler guide that runs across both pages in a spread, double-click in the ruler at the point at which you want the guide. Alternatively, drag from a ruler but release the mouse when the cursor is on the Pasteboard area surrounding the page.

3. To reposition a ruler guide, select the **Selection tool**, position your cursor on an existing guide; then press and drag. A double-headed arrow cursor appears, indicating that you have picked up the guide.

   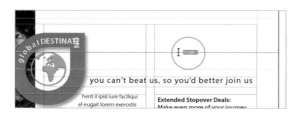

4. To remove a ruler guide, select the **Selection tool** and then click on the guide. It changes color, indicating that it is selected. Press the **Backspace** or **Delete** key. Alternatively, drag the guide back into the ruler it came from.

You can switch **Snap to Guides** off by choosing View > Grids & Guides > Snap to Guides. A tick mark next to the command indicates that **Snap to Guides** is On:

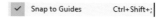

A "spread" usually consists of two pages – a left- and a right-hand page – viewed side by side. For example, when you read a magazine or a book, you are viewing spreads.

Click the right mouse button (Windows), or **Control + Click** (Mac) on the vertical or horizontal ruler to change the ruler units via a context-sensitive menu.

...cont'd

In a document with multiple layers, ruler guides appear on the active layer. When you hide the layer you also hide the layer's ruler guides (see pages 91-92).

Provided that the **Show Transformation Values** checkbox is selected (**Edit/InDesign** > **Preferences** > **Interface**) an on-screen measurements label readout appears at the cursor as you create or move a ruler guide.

Use the keyboard shortcut **Ctrl/Command + Alt + G** to select all guides on a page or spread.

To lock guides on a specific layer, select the layer then choose **Layer Options for ...** from the Layers panel menu (▤), or double-click the layer in the Layers panel and select the **Lock Guides** option.

5 To position a guide with complete precision, drag in a guide and position it roughly where you want it. Click on the guide with the **Selection tool**, then enter the position you want in the **X** or **Y** entry field in the Control panel or the Transform pane of the Properties panel when working in the Essentials workspace. Press **Enter/Return** to apply the new value.

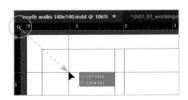

6 To create a vertical and horizontal guide simultaneously, position your cursor in the crosshair area, where the vertical and horizontal rulers meet. Hold down **Ctrl/Command**, then press and drag onto the page. Release the mouse button before you release the **Ctrl/Command** key, otherwise you will reposition the zero point.

Hiding/Showing and Locking/Unlocking Guides

1 To lock an individual guide, click on it with the **Selection tool** to select it. The guide changes color to indicate that it is selected. Then, choose **Object** > **Lock** (**Ctrl/Command + L**). If you try to move the guide, a padlock icon appears indicating the guide's locked status. To unlock a locked guide, click on it to select it, then choose **Object** > **Unlock All on Spread** (**Ctrl/Command + Alt/Option + L**). This option also unlocks any locked objects on the page or spread at the same time.

2 To lock all the guides for an entire document, regardless of which layer they are on, choose **View** > **Grid & Guides** > **Lock Guides** (**Ctrl/Command + Alt/Option + ;**). A tick mark next to the option in the menu indicates that the guides are locked. Use the same option to unlock the guides.

3 To hide/show all ruler guides as well as margin and column guides, choose **View** > **Grids & Guides** > **Hide/Show Guides** (**Ctrl/Command + ;**). This command does not hide frame edge guides, which define the size and position of frames on a page. Choose **View** > **Extras** > **Hide/Show Frame Edges** (**Ctrl + H** – Windows; **Command + Alt/Option + H** – Mac) to hide/show frame edges.

Saving Documents

It is important to be able to use the **Save As**, **Save**, and **Save a Copy** commands as necessary. To avoid losing work, save soon after you start working on a document and remember to save regularly as you make changes.

Save As

Use the **Save As** command soon after starting a new document. **Save As** enables you to specify a folder and name for the document.

1 To save a file for the first time, choose **File** > **Save As** (**Ctrl/ Command + Shift + S**). Use standard Windows/Mac dialog boxes to specify the folder into which you want to save the document.

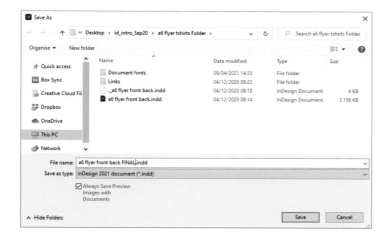

2 Make sure the **File name** entry field is highlighted. Enter a name for

the document. Leave the **Save As Type** pop-up (Windows) or **Format** pop-up (Mac) set to **InDesign 2021 document**. Click on **Save**. The name of the document appears in the document name tab or the title bar of the InDesign document window.

Save

Use the **Save** command regularly as you build your InDesign document so that changes you make are not accidentally lost due to

You can also use the **Save As** command when you want to create a new version of the document on which you are working. Use **File** > **Save As**, specify a different location and/or enter a different name for the file. When you OK the Save As dialog box you continue to work on the new file; the original remains as it was when you last saved it.

Use the shortcut **Ctrl/ Command + Alt/Option + Shift + S** to save all open documents at the same time.

To revert to the last saved version of a file, choose **File** > **Revert**. Confirm the revert in the warning dialog box. The file reverts to the stage it was at when you last used the Save command. This option can sometimes be more efficient than using repeated Undo commands.

...cont'd

any system or power failure. Each time you use the **Save** command, changes you have made to the document are added to the already saved version of the file. You do not need to rename the file or specify its location every time you use the Save command.

To save a file, choose **File** > **Save** (**Ctrl/Command + S**) at regular intervals as you build your document.

Save a Copy

Use the **Save a Copy** command to save a copy of a document at its present state. When you have saved a copy, you continue to work on the original file, not the copy.

To save a copy of a document at its present state, choose **File** > **Save a Copy**. Use standard Windows/Mac dialog boxes to specify the folder into which you want to save the document. Make sure the **File name** entry field is highlighted. Enter a name for the document. Click the **Save** button. The copy is saved in the location and with the name specified. You continue to work on the original file.

Saving Templates

Save a document as a template when you want to create a series of documents with consistent layout, text formatting, color and graphic elements. For example, if you are creating a monthly newsletter, set up an InDesign document that contains all of the standard guides, master pages, style sheets, colors, placeholder frames and graphics. Each time you begin a new issue, open the template and import new content into the existing structure; this will ensure consistency from month to month.

To save a document as a template, follow the steps for the **Save As** command, but choose the

Save as type:	InDesign 2021 document (*.indd)
	InDesign 2021 document (*.indd)
	InDesign 2021 template (*.indt)
	InDesign CS4 or later (IDML) (*.idml)

Format	✓ InDesign 2021 document
	InDesign 2021 template
	InDesign CS4 or later (IDML)

InDesign ... template from the **Save as type** pop-up (Windows) or the **Format** pop-up (Mac).

Choose **File** > **Close** when you finish working on a document. If you have not already saved the file, in the warning box that appears, click **Save** to save and close the file, click **Cancel** to return to the document without saving, or click **Don't Save** to close the document without saving any changes.

When saving InDesign files, avoid using **reserved characters** such as: / \ : ; * ? < > , $ %. Reserved characters can have special meanings in some operating systems and can cause problems when files are used on different platforms.

2 Building Pages

This chapter shows you how to create, manipulate and control text and graphic frames, lines and basic shapes to achieve the exact page layout structure you require.

Creating Frames

The **Frame tools** create containers that define areas on your page that will hold text or images. You can construct the basic layout of a page using frames before you import text and images, or you can create a frame and begin to work with its content immediately.

There are three sets of tools you can use to create frames. The **Rectangle Frame, Ellipse Frame** and **Polygon Frame tools** create frames into which you can place images. The **Rectangle, Ellipse** and **Polygon tools** allow you to create shape frames. Typically, you use shape frames to create simple graphic objects on your page. You can use the **Type tool** to create text frames into which you can type or import text.

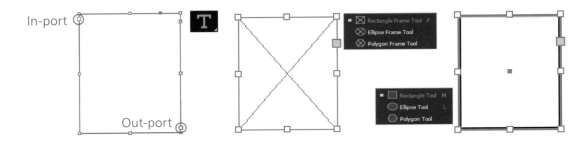

In-port

Out-port

See page 42 for information on working with the Live Corners box

(▢) that appears when you select a rectangle shape or text frame with the **Selection tool**. Ellipses, polygons and stars do not have corner controls.

Empty text frames display In- and Out-ports at the upper-left and lower-right corners, respectively, when selected with the **Selection tool**. Empty graphic frames display an **X** running through the middle (provided that **View > Extras > Show Frame Edges** is selected). Empty shape frames initially have a default 1-point black stroke and no fill.

1 To draw a frame, select the appropriate tool; the cursor changes to the Drawing tool cursor. Press and drag away from the start point, and release when the frame is the size you want. Don't worry if you don't get the shape exactly right to begin with – you can always resize and reposition the frame at a later stage.

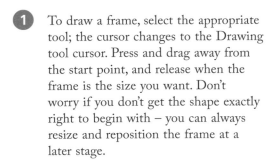

Provided that you do not have a text frame active, you can press the F key on your keyboard to select the **Rectangle Frame tool**.

2 When you release the mouse button, the shape is "selected" – it appears in a "bounding box" that has eight selection handles around the outside. The selection bounding box disappears when you deselect the object. A basic shape frame takes on any fill and/

or stroke attributes currently set in the Tool panel. (See pages 104-105 for information on working with fill and stroke.) Unlike basic shapes, a graphic frame does not take on any fill and/ or stroke attributes currently set in the Tool panel. By default, graphic frames are ready to hold an imported image or graphic.

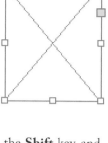

Choose **View > Extras > Hide Frame Edges** to hide the default blue lines that define the size and shape of a frame. Frame edges are visual, on-screen guides. They do not print, but are helpful as you build a layout.

3 To draw a square or circular frame, hold down the **Shift** key, and then press and drag away from the start point. The **Shift** key acts as a constraint on the **Drawing tool**. Make sure you release the mouse button before the **Shift** key, otherwise the constraint is lost. You can also use the **Shift** key with the **Polygon tool** to maintain its proportions.

4 To draw a frame from the center out, hold down the **Alt/Option** key before you start to drag to create the frame.

With any of the basic object **Drawing tools** selected, if you position your cursor on the page, and then **click** the mouse, a dialog box appears that allows you to set exact dimensions for a new object. Click **OK** to create an object with the required dimensions:

5 After you draw an object, you can continue to draw further objects, because the **Drawing tool** remains selected. Make sure you choose the **Selection tool** if you want to make changes to the size or position of an object.

6 To delete a frame, select it with the **Selection tool** and then press the **Backspace** or **Delete** key on your keyboard.

Polygon Tools

Use the **Polygon Frame tool** and the **Polygon tool** to draw regular polygons or stars. Use the **Polygon Settings** dialog box to specify the number of sides for a polygon, or the number of points for a star. Use the **Star Inset** setting to create a star and control the width of the spokes on the star. A **Star Inset** value of zero creates a polygon.

1 To specify the number of sides in a polygon or a star inset to create a star, double-click one of the **Polygon tools**. Enter values as required. **OK** the dialog box, and then press and drag to create a shape using these settings.

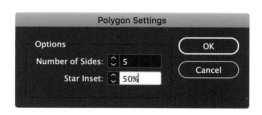

Selection Techniques

An object in InDesign is any shape created with one of the basic shape Drawing tools, any frame created with one of the frame tools, a line created with the **Line tool**, or any path created with the **Pen** or **Pencil tool**. Objects are the fundamental building blocks of all InDesign documents.

If a shape frame does not have a fill you can only select it by clicking on its path (the edge) – it does not select if you click inside the shape.

The **Selection tool** is used to select, resize and move objects or groups. The **Direct Selection tool** is used to edit the shape of paths or frames by working directly on the **Anchor points** that form the shape. The **Direct Selection tool** can also be used for working on the content of graphic frames.

A fundamental technique in any page layout application is that of selecting objects before you make changes to them. In Adobe InDesign, you use the **Selection tool** to select and deselect objects.

① Make sure you have the **Selection tool** selected. Click on an object to select it. With the exception of horizontal or vertical lines, a blue bounding box with eight selection handles appears around the object, indicating that it  is selected. A selection bounding box represents the horizontal and vertical dimensions of an object. The eight selection handles allow you to change the width and/or height of the object. Rectangular graphic and text frames also display a yellow box that allows you to access Live Corner controls (see page 42 for further information). Vertical/ horizontal lines have selection handles at both ends.

② To deselect one or more objects, click into some empty space with the **Selection tool**, or choose **Edit > Deselect All** (**Ctrl/Command + Shift + A**).

③ To select more than one object, click on the first object to select it, hold down the **Shift** key, and then click on additional  objects to add them to the selection. Multiple objects selected in this manner form a temporary grouping: if you move one of the objects, the other selected objects move, maintaining the relative position of each object. This temporary grouping is lost as soon as you select a different object or click in some empty space.

4 Another technique for selecting multiple objects is to "marquee" select them. With the **Selection tool** selected, position your cursor so that it is not touching any object on the page. Press and drag; as you do so, a dotted

marquee box appears. Any object that this marquee touches will be selected when you release the mouse button. This is a very powerful selection technique, and it is worth practicing it a few times to become familiar with it.

5 With any tool selected, you can choose **Edit > Select All (Ctrl/Command + A)** to select all objects on the currently active page or spread, as well as on the Pasteboard area surrounding the page.

6 The **Select** submenu provides useful controls for selecting objects in complex areas of overlapping objects. Select an object, and choose **Edit > Select** to access the options in the sub-menu. (See page 88 for further information on changing the stacking order of objects.)

Select an empty text, graphic or shape frame, then choose **Object > Content** to access the Frame type sub-menu. Choose from **Graphic**, **Text** or **Unassigned** to convert the frame to a different type:

✓	Graphic
	Text
	Unassigned

31

The **Select All command** does not select objects on locked or hidden layers.

First Object Above	Ctrl+Alt+Shift+]
Next Object Above	Ctrl+Alt+]
Next Object Below	Ctrl+Alt+[
Last Object Below	Ctrl+Alt+Shift+[

Moving Objects

You can move objects anywhere you want on the pages of your document or onto the Pasteboard area surrounding your pages. Objects can cross over margins, and you can create a "bleed" by positioning an object so that it runs across the edge of a page onto the Pasteboard.

If you are creating bleeds, make sure that the object extends at least 0.2 in/3 mm beyond the edge of the page to allow for slight misalignment when the page is trimmed. It is a good idea to ask your commercial printer for advice on exactly how much bleed is required.

1 To move or reposition an object, select the **Selection tool**, position your cursor inside the object, and then press and drag to move the object to a new location. If you are moving an image, make sure you don't position your cursor on the Content Grabber circles that appear at the center of the image. (See the box below for further information.)

2 To constrain the movement to vertical, horizontal or increments of 45 degrees, hold down **Shift** and then press and drag to reposition the object. Remember to release the mouse button before you release the **Shift** key, otherwise the constraining effect of the **Shift** key is lost.

Objects placed on the Pasteboard do not print, but they are saved as part of the document and therefore add to the file size. Objects left on the Pasteboard can also add to the processing time when you print. It is a good practice to remove unnecessary objects from the Pasteboard prior to the final output.

3 To move a selected object in increments, press the **Up**, **Down**, **Left** and **Right** arrow keys on the keyboard. Each time you press an arrow key, the object moves .25 mm. To change the nudge increment, choose **Edit > Preferences > Units & Increments** (Windows), or **InDesign > Preferences > Units & Increments** (Mac). Enter a new value in the **Cursor Key** entry field. Hold down **Shift** while pressing an arrow key to move an object 10 times the **Cursor Key** setting.

Content Grabber

When moving images with the **Selection tool**, be careful that you don't drag the "Content Grabber". The Content Grabber appears at the center of the image when you move the **Selection tool** over an image. You can use the Grabber to reposition the image within the frame without first having to select the **Direct Selection tool**. (See page 78 for further information.) The Content Grabber moves the image inside a frame, not the frame itself.

Hold down the **Alt/Option** key and drag an object to create a copy of the object as you drag it. A hollow arrow head appears with the Move cursor to indicate that you are copying:

32

Manually Resizing Objects

Once you've drawn an object, you can resize it manually using the **Selection tool**. You can resize basic objects, open paths, graphic and text frames, and groups.

For basic shapes and paths, if you press and drag on a selection handle in one movement, you see a blue bounding box that represents the new size of the object:

1 To resize a basic shape or frame, select the **Selection tool**. Click on the object; the object is highlighted and eight selection handles appear around the outside. An irregularly shaped object appears within a blue rectangular bounding box with handles.

If you press on the selection handle, pause momentarily and then drag, you see a representation of the complete shape as you resize it:

2 Drag the center left/right handle to increase/decrease only the width of the object. Drag the center top/bottom handle to resize only the height. Drag a corner handle to resize width and height simultaneously.

When you use this technique on an object with Text Wrap applied to it, you see a live preview of text reflowing as you change the size of the object.

3 To maintain the proportions of an object or group as you resize it, hold down **Shift** and drag a selection handle.

4 To resize a line drawn at an angle, select it using the **Selection tool**. Notice that dragging a handle changes the start or end position of the line, in effect changing the length of the line. To change the thickness of a line, use the Stroke panel. (See pages 104-105 for further information on using the Stroke panel.)

When you resize an object, an on-screen readout of Width and Height dimensions appears at the cursor as you drag, making it easy to resize objects to precise dimensions:

5 When you manually resize an object using the **Selection tool**, the Stroke weight remains constant.

W: 9p0
H: 9p0

Control and Transform Panels

When you need to manipulate an object with numerical accuracy you can use the Control panel, the Transform panel or the Transform pane of the Properties panel. When you select an object with the **Selection tool**, these panels display a range of controls for manipulating it.

The matrix of small squares to the left of the panels are the proxy reference points. Each point refers to a corresponding handle on the bounding box of the object. There is also a reference point for the center of the object. Click on a reference point to specify the point around which the transformation takes place.

The **Control panel** is not a default panel in the Essentials, workspace. Choose **Window** > **Control** to show it if necessary. (See pages 20–21 for information on workspaces.)

The appearance of the Control panel changes when you work in a text frame with the **Type tool** selected. See Chapters 4 and 5 for information on using the Control panel to format type.

The **default zero point** for a page is the top-left corner. Step 1 assumes that the default zero point has not been changed.

A **Constrain Proportions button** is selected when it appears as an unbroken chain in a box. It appears as a broken chain when it is deselected. It is important that you can distinguish between the two states. Click the button to toggle between states:

1 The **X** and **Y** entry fields allow you to position an object precisely. The **X** value specifies the position

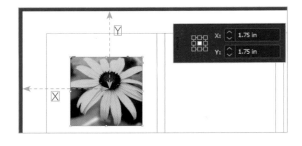

of the object's reference point from the left edge of the page or spread. The **Y** value specifies the position of the object's reference point from the top of the page.

2 To scale an empty text, graphic or shape object to exact dimensions, enter precise values in the **W/H** entry fields. Provided that the **Constrain Proportions** icon is selected, when you change either the **Width** or **Height** value, the other value updates automatically to scale the object in proportion.

3 To scale a frame and its contents (picture or text) as a percentage, enter a scale value in the **Percentage entry** fields, or use the pop-up to choose a preset value. To scale the width, use the **Scale X Percentage** field; to scale the height, use the **Scale Y Percentage** field. You can also enter precise amounts in the **Scale X/Y Percentage** fields (e.g. 12p6). Provided that the **Constrain**

Proportions icon is selected, when you change either the **Width** or **Height** value, the other value

updates automatically to scale the object in proportion.

 Select an object using the **Selection tool** before you make any changes to the fields in the Control or Transform panels, so that settings you create are applied to this object.

④ To rotate a selected object manually, working with the **Selection tool**, position your cursor fractionally outside one of the corner handles. When the Rotate cursor (↷) appears, press and drag in a circular direction to rotate the object. Hold down the **Shift** key as you rotate the object to constrain the rotation to increments of 45 degrees.

⑤ To rotate an object numerically, enter a value in the **Rotate entry** field, or use the pop-up to choose from the preset list. You can enter a positive or negative number from 0 to 360 degrees. You can also click the **Rotate 90° Clockwise/ Counter-clockwise** buttons to rotate in convenient preset amounts.

 Press **Enter/ Return** to apply any new value you enter in the Control or Transform panel.

35

⑥ To apply a stroke to an object, click into the **Stroke Weight** field and then enter a value, or choose a value from the **Stroke Weight** list. Choose a style for the stroke from the **Stroke Type** pop-up list.

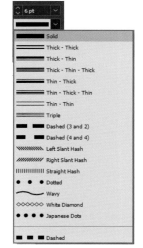

	Solid
	Thick - Thick
	Thick - Thin
	Thick - Thin - Thick
	Thin - Thick
	Thin - Thick - Thin
	Thin - Thin
	Triple
	Dashed (3 and 2)
	Dashed (4 and 4)
	Left Slant Hash
	Right Slant Hash
	Straight Hash
	Dotted
	Wavy
	White Diamond
	Japanese Dots
	Dashed

To correct a mistake, choose **Edit** > **Undo** (**Ctrl/Command** + **Z**). The Undo command is dimmed if you cannot undo an operation. Choose **Edit** > **Redo** (**Ctrl/ Command** + **Shift** + **Z**) to reverse through any undos.

...cont'd

Hot tip

The **Measure tool** (K), in conjunction with the Info panel, is useful when you need to measure the distance from one object to another. Select the **Measure tool**, then drag from one point to another. When you release the mouse button, a measure line appears on your screen. A readout of the distance appears in the Info panel:

Select the **Selection tool** to hide the measure line.

The **Transform panel** is a convenient floating panel that provides a range of controls similar to, but less comprehensive than, those in the Control panel. Choose **Window > Object & Layout > Transform** to show the panel:

7 To shear an object, enter a value in the **Shear entry** field, or use the pop-up (■) to choose from the preset list.

8 When you select a graphic frame that contains an image, click the **Select Container** or **Select Content** button to toggle the selection status between the frame and the image inside.

9 Select an object using the **Selection tool**, then use the **Select Next/Previous Object** buttons to cycle through and select objects on the page or spread. Hold down **Shift**, then click a button to skip by 5 through objects on a page or spread.

10 Use the **Horizontal Flip** and **Vertical Flip** buttons to quickly flip objects. Check which reference point is selected in the Control panel before you flip an object so that you don't end up with unexpected results. The examples below use the center reference point. The flip indicator to the right of the **Flip** buttons provides an excellent visual check for any flip setting applied to an object.

Fill and Stroke pop-up panels

You can quickly access the Swatches panel as a pop-up panel from the Control panel. Click either the **Fill** or **Stroke** triangle to apply color as required. (See pages 104-105 for further information on working with Fill and Stroke.)

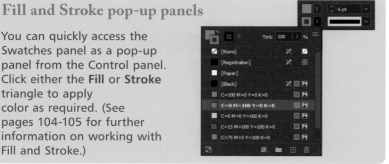

Lines

Use the **Line tool** to create horizontal and vertical lines, or lines at any angle. You can modify lines using the **Selection tool** or using the Control or Transform panels. Use the Stroke panel to change line weight, add arrow heads, and create dashed or dotted lines.

1 To draw a horizontal or vertical line, select the **Line tool**. Position your cursor on the page, hold down the **Shift** key, and then press and drag. Release the mouse button before the **Shift** key when the line is the desired length. The line remains selected when you release, indicated by the two selection handles.

2 To edit a horizontal or vertical line, select the **Selection tool**, click on the line to select it, and drag one of the end points. The vertical/horizontal constraint remains in effect as you resize the line.

3 To draw a line at any angle across your page, select the **Line tool**, position your cursor on the page, and then press and drag. When you release, the line remains selected, indicated by a selection bounding box with eight selection handles.

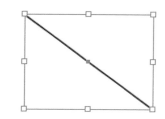

4 To edit a line at any angle, select the **Selection tool** and then click the line to select it. The bounding box appears around the line. Drag any of the selection handles to change the start or end point of the line. Alternatively, select the line with the **Direct Selection tool** (▶). Drag an end point to resize the line.

5 To change the thickness of a selected line, choose **Window > Stroke** (or click the **Stroke** icon in the Panel Dock if you are working with either the Advanced or Essentials Classic workspace) to show the Stroke panel. To change the thickness of the line, you can click the **Weight increment** buttons ( ) to change the thickness in single-point increments, enter an exact value in the **Weight entry** box and press **Enter/Return**, or choose a setting from the Weight pop-up list.

 For any Drawing tool, provided that View > Grids & Guides > Snap to Guides is selected, the Drawing tool cursor snaps or locks onto a guide when it comes within four screen pixels of a column, margin or ruler guide. For graphic and shape frame tools, the cursor displays a hollow arrow head to indicate when it is snapping to a guide. Snap to Guides is useful when you need to create and position objects with precision.

 To convert a horizontal or vertical line to an angled line, select it with the Direct Selection tool and then drag one of the end Anchor points:

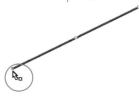

In the Appearance pane of the Properties panel (Essentials workspace), click Stroke to access the full set of Stroke panel controls:

...cont'd

Lines and the Control Panel

You can use the Control panel to control line attributes with numeric precision.

You can drag using the **Pencil tool** to draw freeform lines. The **Smooth tool**, in the **Pencil tool group**, allows you to make a path smoother by dragging across it. The **Erase tool** allows you to erase portions of a selected path by dragging across it.

1 Use the **Selection tool** to select a line you want to change. Click on the start, end or midpoint reference point for the line, to specify the point on the line to which changes refer.

2 Enter values in the **X** and **Y** entry fields to specify the exact position for the chosen reference point of the line.

See page 34 for information on working with coordinates.

3 Enter a value in the **L** entry field to specify the length of the line. Alternatively, use the **Scale X Percentage** and the **Scale Y Percentage** controls to change the length and thickness of the line; **Scale X** and **Scale Y** have the same effect on a line, provided that the **Constrain Proportions for width & height** button is selected.

4 Use the **Rotate** and **Shear** entry fields to rotate and shear the line. Shearing on a line becomes apparent when you increase its stroke weight. (See pages 170-171 and 173 for further information on rotating and shearing objects.)

5 To apply a stroke to an object, click into the **Stroke Weight** field and then enter a value, or choose a value from the Stroke Weight list. Choose a style for the stroke from the **Stroke Type** pop-up list.

You can use the **Transform panel** to change line attributes with numeric precision. Choose **Window** > **Object & Layout** > **Transform** to show the panel. The controls available in the panel offer a convenient subset of those in the Control panel.

Stroke pop-up panel

To change the stroke color for a selected object using the Control panel, click the Stroke pop-up triangle to apply color as required. (See page 106 for further information on working with the Swatches panel.)

Cut, Copy, Paste, Clear

The Clipboard provides one of the most convenient and flexible methods for copying, cutting and pasting objects, frames, groups and text. The limitation of the Clipboard is that it holds the result of only one Cut or Copy command at a time. As soon as you perform another Cut or Copy operation, the newly cut or copied object replaces the previous contents of the Clipboard.

The **Clipboard** is a temporary, invisible storage area. If you cut an important object to the Clipboard, paste it back into the document as soon as possible to minimize the risk of accidentally overwriting it with another Cut or Copy command.

1 Use the **Selection tool** to select an object, frame or group, then choose **Edit > Cut/Copy**. Copy leaves the original on the page and places a copy of the selected object onto the Clipboard. Cut removes the selected object from the page and places it on the Clipboard.

Cut	Ctrl+X
Copy	Ctrl+C
Paste	Ctrl+V
Paste without Formatting	Ctrl+Shift+V
Paste Into	Ctrl+Alt+V
Paste in Place	Ctrl+Alt+Shift+V
Clear	

2 Highlight a range of text using the **Type tool** and then choose **Edit > Cut/Copy** to place selected text onto the Clipboard. (See page 47 for further information on cutting, copying and pasting text.)

The contents of the Clipboard are not saved when you exit InDesign.

3 To paste objects and frames that have been copied to the Clipboard back into a document, make sure the **Selection tool** is selected, and choose **Edit > Paste (Ctrl/Command + V)**. The contents of the Clipboard are pasted into the center of your screen display.

4 Choose **Edit > Clear (Backspace)** for a selected object, frame or group to delete the selected object(s) from the document completely. The clear command does not use the Clipboard and therefore does not affect its contents.

After you cut or copy an object or text to the Clipboard, you can paste it any number of times into your document or into any other publication.

5 Choose **Edit > Paste in Place (Ctrl/Command + Alt/Option + Shift + V)** to paste an object or group onto a page in exactly the same position as that from which it was cut.

Paste without Formatting

Use the **Paste without Formatting** command when you have copied/cut text that already has formatting applied and you want to paste it into a new location without importing the existing source formatting – in other words, pasting into the destination and adopting the destination text formatting.

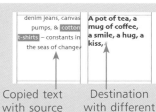

Copied text with source formatting

Destination with different formatting

Result

Smart Guides

Hot tip

Smart Guides is switched on by default. Smart Guides are temporary guides that appear as you drag to create objects and as you move and resize objects. They disappear when you release the mouse button.

Smart Guides appear automatically when you draw, resize, move and align objects on your page as you perform actions to build a layout. The benefit of working with Smart Guides is that you can work accurately from the start, with less fine-tuning to be done at a later stage. Smart Guides can also reduce the need to access dialog boxes and panels to achieve accurate and consistent positioning of objects.

1 To hide/show Smart Guides, choose **View** > **Grids & Guides** > **Smart Guides** (**Ctrl/Command + U**).

2 To control which types of Smart Guides appear when Smart Guides is switched on, choose **Edit** (Windows)/**InDesign** (Mac) > **Preferences** > **Guides & Pasteboard**. In the **Smart Guide Options** area, deselect **Smart Guide checkboxes** for any categories you don't want to appear.

Drawing Objects Using Smart Cursors/Smart Guides

When you draw objects using tools such as the **Rectangle Frame tool**, the **Rectangle tool**, the **Line tool**, the **Pen tool** and the **Type tool**, you can use Smart Cursors to draw new objects aligned to existing objects on the page. Also, Smart Guides display to indicate matching dimensions for objects.

Don't forget

Typically, Smart Guides indicate when objects align with the edge or center of another object or page.

1 To draw an object aligned to another object, select one of the object **Drawing tools**. As you position your cursor to draw the new object, watch the Drawing cursor carefully. The appearance of a white triangle at the cursor indicates that the cursor aligns with an edge or the center of an existing object. Drag with the **Drawing tool**. The start of the new object aligns accurately with the edge of the existing object:

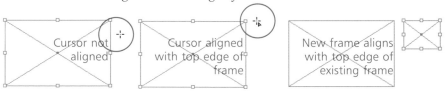

Cursor not aligned | Cursor aligned with top edge of frame | New frame aligns with top edge of existing frame

2 To draw a frame to match the size of an existing frame, working with a frame tool or the **Type tool**, position your cursor so that it is aligned with an edge of the existing frame. (See previous technique.) Start to drag to define the size of the new frame. Vertical and horizontal Smart Dimension guides appear, to indicate that heights/widths match when the cursor reaches the correct position.

To switch Smart Guides On and Off you can also use the Smart Guides option in the View Options pop-up menu, located near the bottom of the tool bar:

Positioning Objects Using Smart Guides

Smart Guides make it quick and easy to position and align an object to the center or edge of another other object, or the page. You can also create equal spacing between objects using Smart Spacing.

1 Make sure that Smart Guides is switched on. Using the **Selection tool**, start to drag an object across your page. Various Smart Spacing and Align guides appear as you reposition the object, to indicate where objects align. In this screenshot, arrows on the right indicate equal vertical

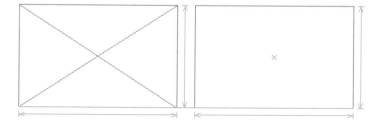

spacing between the three objects. The vertical guide indicates alignment of the edge of the reversed-out text with the left edge of the paragraph.

2 Light magenta Smart Guides appear to indicate alignment to the vertical and/or horizontal center of the page. In this example, the top and right edges of the graphic frame align to the center of the page.

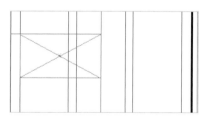

Dark magenta Smart Guides indicate the horizontal center of a column or gutter.

Live Corners

You can switch off Live Corner controls on frames by choosing View > Extras > Hide Live Corners:

When you draw a rectangular frame using the **Rectangle Frame tool**, the **Rectangle tool** or the **Text tool**, then select it with the **Selection tool**, a yellow square appears near the top right of the frame edge. You can use the yellow square to access "Live Corners" – a functionality that allows you to change the appearance of corners for the selected rectangle or square interactively by dragging with the mouse.

1 To access Live Corners, select a basic rectangle using the **Selection tool**, then click the yellow square. This activates four yellow diamond markers at the corners.

2 To change the corner radius for all corners, simply drag one of the yellow diamond corner markers.

3 To change the corner radius for an individual corner, hold down **Shift**, then drag the yellow diamond marker for the corner you want to change.

4 To cycle through different corner effect types, hold down **Alt/Option** then click a yellow diamond marker. To change the corner effect for one corner only, hold down **Alt/Option + Shift** and click the diamond marker for the corner you want to change.

5 To hide Live Corner controls, simply click on some white space away from the selected object.

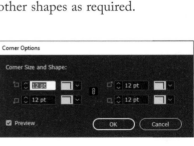

Corner Options

You can also control the appearance of corners on rectangles using the **Corner Options** dialog box. An advantage of using the dialog box is that you can specify exact values for the size of an effect, making it easy to replicate the same effect on other shapes as required.

In the **Corner Options** dialog box, keep the **Make all settings the same** button (🔲) selected to apply changes consistently to all corners of the rectangle. Switch it **Off** (🔲) to create settings for corners individually.

1 Select a rectangle using the **Selection tool**, then choose **Object** > **Corner Options**.

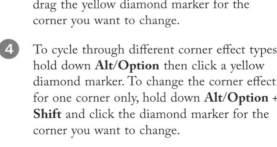

3 Text Basics

This chapter covers essential techniques such as importing text, highlighting text, basic text editing, and threading or linking text frames. It is important to master these everyday procedures.

Entering and Importing Text

44

You cannot position the **Type tool cursor** inside an existing text frame to draw another frame.

There is a range of techniques for entering text into an InDesign document. You can enter text directly into a text frame using the keyboard, paste text already stored on the Clipboard into a text frame, or import a text file prepared in a wide variety of word processing applications into a text frame or directly into the InDesign document. Remember – when you type, paste or import text, it appears at the Text Insertion Point.

1. To create a text frame, select the **Type tool**. Position your cursor on the page, then press and drag. This defines the size and position of your text frame. When you release, you will see a Text Insertion Point flashing in the top-left corner of the frame.

2. To enter text directly into the frame, begin typing on the keyboard. Text wraps automatically when it reaches the right edge of the text frame. Press **Enter/Return** only when you want to begin a new paragraph. By default, text is formatted as **Minion Pro Regular 12pt**.

Begin typing on the keyboard to enter text ...|

3. To import a word processed file, make sure you have a text frame, a graphic frame or a basic shape selected. You need not select the **Type tool**. Choose **File >**

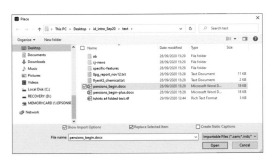

If the text file you import contains more text than will fit into the text frame, the **overflow marker** appears near the bottom-right corner of the frame to indicate that there is **overmatter**:

Once upon a time, there was a poor hunter. One day he came across a trapped crane. He took pity on the crane and released it. A few days later, a lovely woman visited his house, and asked him for shelter for

You can make the frame bigger, or the text smaller, or you can thread the text into another frame (see pages 49-50) to make the remaining text visible.

Place. Use standard Windows/Mac dialog boxes to locate the text file you want to place. Click on its name to select it. If you want to make sure that typographer's quotes (rather than foot and inch symbols) are used in the imported text, or if you want to remove formatting already applied in the Word file, then select **Show Import Options**. When you click **Open**, a secondary **Import Options** dialog box appears. Select the **Use Typographer's Quotes** option to ensure that straight quotes and apostrophes are converted into typographic equivalents. Select the **Remove Styles and Formatting from Text and Tables** radio button to discard Word formatting.

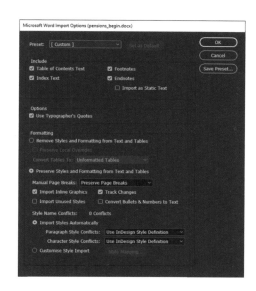

4 Click the **OK** button. Graphic frames and basic shapes are converted automatically into text frames, and the text flows in.

Mapping Word Styles

The **Microsoft Word Import Options** dialog box allows you to map styles from a Word document to Paragraph Styles in InDesign.

1 In the **Import Options** dialog box, select the **Preserve Styles and Formatting from Text and Tables** radio button. Select the **Customize Style Import** radio button and then click **Style Mapping**.

2 In the **Style Mapping** dialog box, select a Word style, then select the InDesign paragraph style you want to map it to from the pop-up menu that appears in the InDesign Style column. Click **OK**.

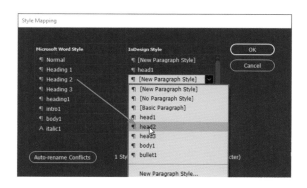

Hot tip

To place text without first creating a text frame, use the Place command to locate the text file you want to place. After you select a file and click **Open**, the cursor changes to the **Loaded text cursor**:

Position the cursor, then click to place the text into a default-width text frame. If you click within column guides, InDesign creates a text frame to the width of the guides. Alternatively, press and drag with the Loaded text cursor to define the width and height of the text frame into which the text will flow.

45

Hot tip

With the Text Insertion Point in a text frame, you can quickly fill the entire frame with placeholder text by choosing **Type > Fill with Placeholder Text**. InDesign inserts meaningless Latin text to fill the frame.

Basic Text Editing

Almost as soon as you start working with text, you will need to be able to make changes and corrections – correcting spelling errors, changing punctuation, deleting words, and so on. Use the **Type tool** to make changes to text.

Hot tip

If you are working with the **Selection tool**, you can double-click on a text frame to quickly select the **Type tool** and place the Text Insertion Point where you click. Press the **Esc** key to quickly reselect the **Selection tool**.

1. To make an alteration to your text, select the **Type tool**. Click in the text frame where you want to make changes. The text frame does not have to be selected before you click into it; if the frame is selected, the selection handles on the frame disappear. A Text Insertion Point appears at the exact point at which you click. The Text Insertion Point is the point at which text will appear if you type it on the keyboard, paste it from the Clipboard, or import it using the **Place** command.

> Once upon a time, there was a poor hunter. One day he came across a trapped|crane. He took pity on the crane and released it. A few days later, a lovely woman visited his house, and asked him for shelter for the night.#

2. Notice that when you move your cursor within an active text box, it becomes an I-beam cursor. Position the I-beam cursor anywhere in the text and click to reposition the Text Insertion Point.

3. To delete one character to the left of the Text Insertion Point, press the **Backspace** key. To delete one character to the right of the Text Insertion Point, press the **Delete** key.

Hot tip

To show/hide invisible text characters such as spaces, tabs and carriage return markers, choose **Type > Show/Hide Hidden Characters**. The hashmark symbol (#) represents the end of the text file:

> And so on¶
> Once upon a time, there was a poor hunter. One day he came across a trapped crane.¶
> »   He took pity on the crane and released it. A few days later, a lovely woman visited his house, and asked him for shelter for the night.#

The Text Insertion Point

Use the following techniques for fast and efficient editing of text in your documents.

1. To move the Text Insertion Point character by character through the text, press the **Left** or **Right** arrow key on the keyboard. To move the cursor up or down one line at a time, press the **Up** or **Down** arrow key.

2. To move the Text Insertion Point one word left or right at a time, hold down **Ctrl/Command** and use the **Left/Right** arrow keys.

...cont'd

3 To move the Text Insertion Point up or down one paragraph at a time, hold down **Ctrl/Command** and use the **Up/Down** arrow keys.

4 To move the Text Insertion Point to the end of a line, press the **End** key; to the beginning of a line, press **Home**.

5 To move the Text Insertion Point to the start of the story, hold down **Ctrl/Command** and press the **Home** key. To move the Text Insertion Point to the end of the story, hold down **Ctrl/Command** and press the **End** key.

Select the **Type tool** and then click into an empty graphic or shape frame to convert it into a text frame.

Cutting, Copying and Pasting Text

Use the **Cut**, **Copy** and **Paste** commands to move words, phrases or paragraphs from one place to another in a document. You can also use these commands to copy text from one InDesign document to another, or from Microsoft Word into InDesign.

1 To copy or cut text, first highlight a range of text. Choose **Edit > Cut/Copy** to store the text temporarily on the Clipboard.

Add **Shift** to any of the cursor movement keyboard combinations to highlight a range of text that the Text Insertion Point moves across. For example, to highlight from the current Text Insertion Point to the end of the text file, including any overmatter, hold down **Ctrl/Command + Shift** and press the **End** key.

2 To paste text back into a text frame, select the **Type tool**, then click into a text frame to place the Text Insertion Point

Edit	
Undo Move Item	Ctrl+Z
Redo Move Item	Ctrl+Shift+Z
Cut	Ctrl+X
Copy	Ctrl+C
Paste	Ctrl+V
Paste without Formatting	Ctrl+Shift+V
Paste Into	Ctrl+Alt+V
Paste in Place	Ctrl+Alt+Shift+V
Clear	Backspace

at the point where you want to place the text. Choose **Edit > Paste**.

3 Choose **Edit > Clear**, or press the **Backspace** key on your keyboard, when you want to delete selected text without storing a copy of it on the Clipboard.

If you choose **Edit > Paste** without first positioning the Text Insertion Point, a copy of the text on the Clipboard is placed in a new text frame in the center of the screen display.

4 Use the **Paste without Formatting** command to paste text without any of the formatting currently applied to it. Using this command, text takes on the formatting of the text where it is pasted.

Highlighting Text

Use the **Type tool** to select text. Once you have highlighted text you can delete, overtype, cut or copy it. Highlighting text is also a crucial step before you change its formatting.

You sometimes need to zoom in on text in order to highlight the exact characters you want.

1 Position the Type cursor at the start of the text you want to highlight. Press and drag across the text. As you do so the text will "reverse out" to indicate exactly what is selected. Drag the cursor horizontally, vertically or diagonally, depending on the range of text you want to highlight. You must select all the text you want to select with one movement of the mouse: you can't release and then drag again to add to the original selection. Use this technique to select any amount of visible text.

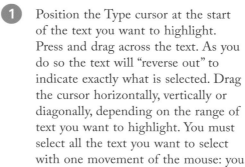

2 Position your cursor on a word, and double-click to highlight one word. This is useful when you want to delete a complete word, or when you want to replace a word with another word by overtyping it.

When you have selected a range of text, if you press any key on the keyboard, you are overtyping the selected text. Whatever you type replaces the selected text. If this happens unintentionally, choose **Edit** > **Undo Typing** immediately.

3 Click three times to select a line.

4 Click four times to select an entire paragraph.

5 Position your cursor at the start of the text you want to highlight. Click the mouse button to place the Text Insertion Point. This marks the start of the text you want to highlight. Move your cursor to the end of the text you want to highlight. Do not press and drag the mouse at this stage: simply find the end of the text you want to highlight. Hold down **Shift**, and then click to indicate the end of the text. Text between the initial click and the **Shift + Click** operation is highlighted. This is a useful technique for highlighting a range of text that runs across several pages.

To deselect text, with the **Type tool** selected, click anywhere within the text frame.

6 Click into the text, then choose **Edit** > **Select All** (**Ctrl/ Command** + **A**) to select the entire text file, even if it is linked through multiple pages. This includes any overmatter, even though you cannot see it.

Threading Text

Threading is a technique by which you link text from one frame into another frame when there is too much text to fit in the first frame (and you don't want to make the type smaller or the frame bigger). Additional text that will not fit in a text frame is referred to as overmatter or overset text. It is indicated by a red "+" symbol (⊞) in the Out-port, near the bottom-right corner of the text frame.

You can thread text from one frame to another frame on either the same page or a different page. If necessary, threads can jump multiple pages.

Although you can thread empty text frames, it is easier to understand how threading works if you are flowing a text file.

1 Place a text file into a text frame. (See pages 44-45 for information on placing text files.) Make sure there is more text than will fit into the frame. A red "+" symbol appears in the Out-port.

2 Draw another text frame. Using the **Selection tool**, reselect the frame that holds the text. Click once on the ⊞ symbol. The cursor changes to the Loaded text cursor.

3 Position your cursor in the next text frame. Notice the chain icon that appears in the top-left corner of the cursor to indicate that you are about to thread text into the frame.

4 Click to flow text into the frame.

A text frame drawn with the **Type tool** has empty In- and **Out-ports**, which are visible when you select the frame with the **Selection tool**. Graphic or Shape frames can display In- and Out-ports only when they have been converted to text frames (**Object** > **Content** > **Text**).

The **Info panel** provides character, word, line and paragraph count information when you are working with the **Type tool** in a text frame. A "+" symbol indicates amounts of overset text:

...cont'd

5 Repeat the process until there is no more text to thread. You know that there is no more text to thread when the Out-port is empty:

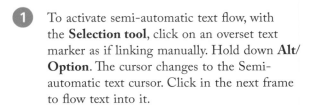

6 You don't have to create another text frame in order to thread text. You can simply click on the ⊞ symbol, position your Loaded text cursor, and then press and drag to define the next text frame. When you release the mouse button, the overset text flows into the area you define.

7 Alternatively, you can position your Loaded text cursor on your page, and then click to place the overset text. InDesign automatically creates a text frame to the width of the page margins, or any column guides within which you click. The frame starts where you click and ends on the bottom margin of the page.

Semi-Automatic Text Flow

Semi-automatic text flow is useful when you have a long text file that you want to thread through several existing text frames.

1 To activate semi-automatic text flow, with the **Selection tool**, click on an overset text marker as if linking manually. Hold down **Alt/Option**. The cursor changes to the Semi-automatic text cursor. Click in the next frame to flow text into it.

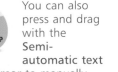

2 Keep the **Alt/Option** key held down. The text cursor reloads if there is more overmatter to place. Click into the next frame. Repeat the process as necessary. Select the **Selection tool** to end the process.

If you place the Loaded text cursor over a text frame that contains text, and then click, the existing text is retained and appears after the threaded text.

You can also press and drag with the **Semi-automatic text flow cursor** to manually define the size of the text frame you want to thread text into. Keep the **Alt/Option** key held down to continue the semi-automatic threading process if required.

Understanding Text Threads

Understanding the symbols that appear in the In- and Out-ports, and how to control and manipulate them, is essential for working efficiently and accurately with threaded frames. Each port is represented by a small square located on the edge of a text frame.

An empty In-port in the top-left corner of a text frame indicates that this is the first frame in the thread – no text flows into this frame from another frame.

A solid triangle symbol in the In-port indicates that the frame is part of a threaded story – text from a preceding frame flows into it.

A red "+" symbol ( ) in the Out-port in the bottom right of a frame indicates overset text. This overset text can be linked to another frame. A solid triangle symbol in the Out-port indicates that this frame is already linked to another frame.

An empty Out-port indicates there is no overset text and that the frame is not linked to any further frames.

Text Threads

Each text thread is represented by a blue line running from an Out-port to an In-port.

1. To show text threads, choose **View > Extras > Show Text Threads**. Select a text frame to see the text threads, which indicate the flow of text through the threaded frames.

Use the keyboard shortcut **Ctrl/ Command + Alt/Option + Y** to show/hide threads. Threads are visible provided that at least one text frame in the threaded story is selected.

51

If you switch off the Show Thumbnails on Place checkbox preference (**Edit/InDesign > Preferences > Interface**), the shape of the Loaded text cursor changes, depending on where you place it. If you place the cursor over empty space on your page or on the Pasteboard, it displays as the **Loaded text cursor**:

If you place the cursor over an empty text frame, it displays as the **Thread text cursor**:

If you place it over an empty graphic or shape frame, it becomes the **Convert to text frame cursor**:

...cont'd

Breaking Threads

In order to manage and control threaded text in a document, you sometimes need to break the thread between connected frames.

If at any time you pick up the Loaded text cursor accidentally, you can "unload" it, without placing or losing any text, by pressing the **Esc** key or clicking on any other tool in the Tool panel.

1 To break the threaded link between frames, select the **Selection tool**, and then click into a frame in a series of threaded text frames. The In- and Out-ports show when you activate a series of threaded frames. Text threads appear, provided that **View > Extras > Show Text Threads** is selected.

2 Click once in the Out-port of the frame where you want to break the thread.

Position your cursor within the frame whose Out-port you clicked. The cursor changes to the Broken Chain text cursor.

3 Click in the frame to break the thread between the frames.

4 As an alternative to Steps 2-3 above, you can double-click an In- or Out-port to break the connection between frames.

You can also break a thread by clicking in the In-port of a threaded text frame, and then clicking anywhere inside the same text frame.

Deleting Frames

1 To delete a frame that is part of a threaded story, select the frame using the **Selection tool**, and then press the **Backspace** or **Delete** key. Text reflows through the remaining frames and no text is lost.

Glyphs

A glyph is a specific instance or form of a character. You can use the Glyphs panel to locate additional characters that are not available on the keyboard. Open Type fonts offer glyphs that are alternative letterforms for the same character; for example, Adobe Caslon Pro has small cap and ornament variations for the letter "A".

Use the **Show pop-up** menu to limit the display of glyphs to a specific subset; e.g. Punctuation:

1 To insert a glyph, using the **Type tool**, click to place the Text Insertion Point in your text. Choose **Type > Glyphs**.

2 In the Glyphs panel, use the Font pop-up at the bottom of the panel to choose a font family. Choose a style from the **Style** pop-up menu next to it.

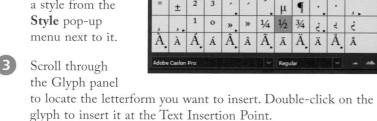

3 Scroll through the Glyph panel to locate the letterform you want to insert. Double-click on the glyph to insert it at the Text Insertion Point.

4 For convenience, the 35 most recently used glyphs appear at the top of the panel. To clear recently used glyphs, position your cursor on a recent glyph, then right-click (Windows), or **Control + Click** (Mac) to access commands in the context menu.

> Delete Glyph from Recently Used
> Clear All Recently Used
>
> Load Glyph in Find
> Load Glyph in Change

Alternative glyphs are not available in all fonts.

5 A small triangle on the glyph box indicates that there are alternative glyphs available. To insert an alternative glyph, press and hold the glyph to reveal the alternatives in a pop-up box. Move your cursor onto the glyph you want to insert, and then release the mouse.

6 Click the **Open Type** button that appears at the bottom right of a text frame selected with the Selection tool, to quickly access any available additional Type options such as Discretionary Ligatures or Small Caps.

Understanding Text Threads

> ☑ Discretionary Ligatures
> Understanding Text Threads

Special Characters and White Space

A **discretionary hyphen** is a hyphen you use to hyphenate words manually. Unlike an ordinary hyphen, the discretionary hyphen will not appear if text reflows and the word no longer needs to be hyphenated.

InDesign makes it easy to insert special characters such as Em- and En-dashes, bullets and discretionary hyphens. You can also insert a range of different spaces to meet specific requirements. Use either the Type menu or a context menu to access these options.

Special Characters

1 To enter a special character, select the **Type tool**, and then click in the text at the point where you want to insert the character.

2 Choose **Type > Insert Special Character**. Select a category from the sub-menu, then select the character you want to insert.

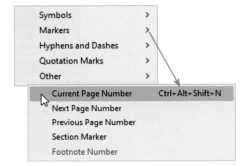

The **Insert Break Character sub-menu** can also be accessed via the Type menu.

3 To use the context menu, right-click (Windows), or **Control + Click** (Mac). Choose **Insert Special Character** from the context menu, select a category, and then click on the special character you want to include in the text.

White Space

InDesign offers a range of white-space options to meet specific typesetting requirements – for example, you might use a Figure space to ensure that numbers align accurately in a financial table.

Forced line breaks are also referred to as "soft returns". They are useful when you want to turn text onto a new line without creating a new paragraph (in headlines, introductory paragraphs, subheads and the like, where it is important to control the look of your text).

The keyboard shortcut is **Shift + Enter/Return**.

1 To insert a space of a specific size, use the techniques described above, but choose **Insert White Space** from the Type or context-sensitive menu to access the White Space sub-menu options.

Em Space	Ctrl+Shift+M
En Space	Ctrl+Shift+N
Nonbreaking Space	Ctrl+Alt+X
Nonbreaking Space (Fixed Width)	
Hair Space	
Sixth Space	
Thin Space	Ctrl+Alt+Shift+M
Quarter Space	
Third Space	
Punctuation Space	
Figure Space	
Flush Space	

Text Frame Options – Columns

The **Text Frame Options** dialog box provides a set of important controls for working with text frames. You can divide a frame into columns and you can control the position of type within frames.

Columns

Dividing a single text frame into a number of columns makes it easy to create multi-column layouts and especially easy to create equal columns that do not necessarily match the number of columns in the underlying page grid – defined during page setup.

Select the **Preview option** in the Text Frame Options dialog box to preview changes in the document before you click the OK button to accept the settings:

1 To specify columns for a text frame, select the frame and choose Object > **Text Frame Options (Ctrl/ Command + B)**.

Click the **Baseline Options** tab in the **Text Frame Options** dialog box to set a custom baseline grid for the selected text frame.

55

2 To maintain the exact dimensions of the text frame but divide it up into equal columns, make sure that the **Fixed Column Width** option is deselected. Enter a value for the Number of columns. Enter a value for the Gutter – the

To specify columns of a specific measure or width, select the **Fixed Column Width option**. Enter values for Number of columns and Gutter, and an exact value for the Width of the columns. The overall width of the text frame is adjusted according to the settings to give the exact number of columns of the specified width.

space between columns. As you change either the Number of columns or the Gutter value, the column Width field changes automatically as InDesign takes into account the size of the text frame and the settings you specify. Using this procedure, the size of the text frame is not altered and you end up with equal columns within the overall frame width.

Text Frame Insets & Vertical Alignment

Insets

Apply insets to a text frame when you have colored the background of the frame or changed its stroke, and you want to move the text inward from the edge of the frame.

Select the **Make all settings the same** button ( ) to apply equal insets to the top, bottom, left and right edges of the frame.

1 To apply text insets to a text frame, select a frame and choose **Object > Text Frame Options**. Enter Inset Spacing values for Top, Bottom, Left and Right. You can use the arrows to change the settings in increments. To specify a value in points, enter a value followed by "pt". When the text frame is selected, the inset is represented by a blue rectangle within the text frame. This disappears when the frame is not selected.

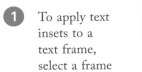

Extended Stopover Deals:
Make even more of your journey with spectacular short breaks in capital cities throughout Europe and Africa.

Extended Stopover Deals:
Make even more of your journey with spectacular short breaks in capital cities throughout Europe and Africa.

Vertical Alignment

The **Vertical Justification** option controls the position of the text vertically in the text frame. The **Center** option is often useful when you have a headline in a color frame.

1 The default Vertical Justification alignment is Top. To change the vertical alignment for text in a frame, choose **Object > Text Frame Options** (**Ctrl/Command + B**). Choose an option from the **Vertical Justification Align** pop-up menu.

4 Character Settings

Setting type is an essential discipline in creating visually attractive publications.

Font, Size, Style

Font

A font is a complete set of characters (uppercase, lowercase, numerals, symbols and punctuation marks) in a particular typeface, size and style – for example, 18 point Arial Bold. The term "typeface" describes the actual design or cut of the specific characters. Arial is a typeface, and there can be many versions of the typeface, such as Arial Narrow, Arial Black, and so on.

Typefaces fall into two main categories – serif and sans serif. Adobe Caslon, used in this paragraph, is an example of a serif typeface. Serifs are the small additional embellishments or strokes that end the horizontal and vertical strokes of a character. Serif faces are often used to suggest classical, established values and traditions.

Gill Sans is an example of a sans serif typeface. Sans serif faces do not have the additional embellishments finishing off horizontal and vertical strokes and are often used to create a modern, contemporary look and feel.

When you choose a font or a set of fonts to use in your document, your aim is to make a choice that reflects the style and identity of the publication, in much the same way that we choose clothes to reflect our personality and identity.

Controls for formatting selected text, available in the **Control panel**, can also be accessed through the **Character panel**. Choose **Type > Character (Ctrl/Command + T)**, to display the floating panel.

In the Essentials workspace, you can use controls in the **Character pane** of the Properties panel:

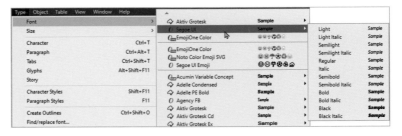

1 To change the font for highlighted text, choose **Type > Font**. Select a font and – when available – a style from the list of available fonts that appears in the sub-menu.

2 Alternatively, make sure that the **Character Formatting Controls** button is selected in the Control panel, and then click on the **Font** pop-up menu (▾) to choose from the font list.

To show the panel menu, click the **panel menu button** (☰) for the panel. The button is located in the top-right corner of the Character panel.

...cont'd

Size

You can enter a type size value from 0.1 to 1296 points. Type size can be entered to .01-point accuracy; for example, you can enter a value of 9.25 points.

1 Make sure you select some text. Choose **Type** > **Size**. Select a size from the preset list in the sub-menu. If you choose "Other" then the size entry field in the Character panel is highlighted. Choosing "Other" will show the Character panel if it is not already showing.

2 Alternatively, in the Control panel, either use the **Size** pop-up to choose from the preset list, or highlight the **Size** entry field and enter a value. Press the **Enter/ Return** key to apply the change. You can also click the arrows ( ) to increase or decrease the point size in 1-pt steps.

Style

1 To change the style – for example, to bold or italic – choose **Type** > **Font**. If a choice of style is available, this is indicated by a pop-up triangle to the right of the typeface name in the font list. If there is no pop-up then there are no style options for the typeface.

2 You can also change the style for

highlighted text using the Style pop-up menu in the Control panel.

3 To apply settings such as **All Caps, Small Caps, Superscript, Subscript, Underline** and **Strikethrough**, click on the appropriate button in the Control panel.

To select and activate fonts from an extensive range of fonts in the cloud-based Adobe Fonts library, click the **Find More option** after you click the **Font** pop-up button:

To increase/ decrease the point size of selected text in 2-point increments, hold down **Ctrl/Command + Shift**, and press the < or > key.

The increment is set in the **Units & Increments preferences** dialog box. Choose **Edit** > **Preferences** > **Units & Increments** (Windows), or **InDesign** > **Preferences** > **Units & Increments** (Mac). Enter a value in the **Size/Leading** entry field to change the increment: an increment of 1 gives greater control and precision.

To manually resize a selected text frame and the text inside it in proportion, select the frame with the **Selection tool**, position your cursor on a corner handle, hold down **Ctrl/Command + Shift**, then drag the handle.

59

Leading

The term "leading" is derived from the days of hot metal typesetting when typesetters would insert strips of lead between lines of type to create additional vertical space.

Leading is a traditional typesetting measurement. It measures the distance from one baseline of type to the next. A baseline is an imaginary line running along the base of type. Leading is an extremely important factor in setting type and can greatly affect the readability of the text on a page.

My beautiful fish

My fish shines in the deep blue ocean waves. My fish shimmers when happy. My fish glows and sparkles when it plays games with his friends. My fish bubbles playfully when he talks. My fish loves the coral reef when he swims past.
My fish sparkles like a rainbow in the sky on a rainy day. My fish is very proud of his scales.

Leading is set relative to the size of the type with which you are working. For example, if your body text size is 10 points, you might set a leading value of 14 points. This is expressed as 10 on 14 (10/14).

To change the default increment of 2pt used by the keyboard shortcut for Leading, choose **Edit > Preferences > Units & Increments** (Windows), or **InDesign > Preferences > Units & Increments** (Mac). Enter a value in the **Size/Leading** field.

When you enter type into a newly created frame, it uses Auto-leading as the default leading method. (See the next page for further information on Auto-leading.)

InDesign applies leading as a character attribute. Enter values for leading from 0 to 5000 points in .01-point accuracy.

Absolute/Fixed Leading

Absolute or Fixed leading uses a fixed value for your leading that does not vary when you change the point size of your type. For example, if you are working with 9-point type with a leading value of 13 points (9/13) and then change the type size to 12, the leading value remains set at 13.

When you are learning InDesign, remember to make use of the **Tool Tips**, which provide explanatory labels for the icons in the Control and Character panels. To show a Tool Tip, position your cursor on an icon; don't move it for a second or so and the Tool Tip should appear:

1 To set a fixed leading value, first select the text to which you want it to apply – the leading value for a line is set according to the highest leading value applied to any character on the line. To set the leading for a paragraph, make sure you select all the characters in the paragraph first.

My beautiful fish

My fish shines in the deep blue ocean waves. My fish shimmers

when happy. My fish glows and sparkles when it plays games with his friends. My fish bubbles playfully when he talks. My fish loves the coral reef when he swims past.

...cont'd

2 Make sure the **Character Formatting Controls** button is selected in the Control panel, or use the Character panel (**Ctrl/Command + T**).

3 Highlight the **Leading** entry field. Enter a leading value. Press **Enter/Return** to apply the value to the selected text.

4 You can also use the **Leading** pop-up to choose from the preset list, or click the arrows ( ) to increase or decrease the leading in 1-point steps.

5 To change leading for selected text using the keyboard, hold down the **Alt/Option** key and press the **Up/Down** arrow keys on the keyboard. The leading changes in 2-point steps.

Auto-leading

Auto-leading sets a leading value equivalent to an additional 20% of the type size with which you are working. When you decrease or increase the point size of your text, the leading will change automatically to a value that is 20% greater than the new point size. You can choose Auto-leading from the **Leading** pop-up in the Control panel, the Character panel or the Character pane of the Properties panel.

1 To set Auto-leading, make sure you have a range of text selected, and then choose **Auto** from the **Leading** pop-up. Auto-leading is represented as a value in brackets.

My beautiful fish
My fish shines in the deep blue ocean waves. My fish shimmers when happy. My fish glows and sparkles when it plays games with his friends. My fish bubbles playfully when he talks. My fish loves the coral reef when he swims past.
My fish sparkles like a rainbow in the sky on a rainy day. My fish is very proud of his scales.

The **Control panel** is visible along the top of your InDesign window when you are working with the Advanced, Essentials Classic, Typography or Book workspaces selected. (See pages 20-21 for information on selecting workspaces.)

You can use the **Eyedropper tool** to copy type settings. Using the **Type tool**, highlight the "target" text whose settings you want to change. Select the **Eyedropper tool**, then click carefully on the "source" text that has the type settings you want to copy. The settings are immediately applied to the target text. Select any other tool in the Toolbox to end the procedure.

To change the default value for **Auto-leading**, first make sure nothing is selected, then in either the Control panel with the **Paragraph Formatting Controls** button selected or the Paragraph panel, choose **Justification** from the panel menu ( ). Enter a new value for Auto-leading, then click **OK**.

Kerning and Tracking

Choose **Edit** >
Preferences >
**Units &
Increments**
(Windows) or
InDesign > **Preferences** >
Units & Increments (Mac),
and enter a value in the
Kerning entry field to
specify the default Kerning/
Tracking increment for the
Kerning or Tracking
keyboard shortcut.

Kerning is a technique of reducing the space between certain pairs of characters that do not produce attractive, graceful results when they occur next to one another, especially at larger point sizes – for example, LA, To, P., WA. In InDesign you can use Manual kerning, Optical kerning or Metrics kerning to achieve balanced, attractive spacing.

Manual Kerning

Manual kerning allows you to kern character pairs in 1/1000th Em units, using the Control panel, the Character panel, or keyboard shortcuts.

1 To manually kern character pairs, first select the **Type tool** and then click to place the Text Insertion Point between two characters.

2 In the Control panel, enter a value in the **Kerning** entry field, and press **Enter/Return** to apply the new value. Alternatively, you can use the **Kerning** pop-up to choose from the preset list, or you can click the arrows (◌) to adjust the kerning in steps of 10. Negative values move characters closer together; positive values move characters further apart.

Use the
shortcut **Ctrl/
Command +
Alt/Option + Q**
to remove all
Manual tracking and
kerning from selected text.
Beware: this shortcut also
resets the kerning method
to Metrics and removes
tracking settings applied
through a paragraph style.

3 Hold down **Alt/Option** and press the **Left** or **Right** arrow key to move characters by the default kerning step of 20/1000th of an Em.

AWAKE
Kerning = 0

4 To remove Manual kerning, make sure your Text Insertion Point is flashing between the character pair, then enter a zero in the **Kerning** entry field. Choose **Metrics** or

AWAKE
Kerning = -110

Optical from the pop-up menu to revert to the in-built value for that kerning method.

Metrics and Optical Kerning

Use **Metrics** as the kerning method when you want to use the pair-kerning information built into a font. **Metrics** is the default kerning method applied when you enter text into a new text frame. The **Optical** method kerns character pairs visually, and can provide good results for type consisting of mixed font and size settings.

1 To apply **Metrics** or **Optical** kerning, select the text and choose an option from the **Kerning** pop-up in the Control panel, Character panel or the Character pane of the Properties panel. A kerning value enclosed in brackets, when your Text Insertion Point is flashing between character pairs, indicates the kerning value used by the Metrics/Optical kerning method.

Tracking

Tracking, or "range kerning", is a technique of adding or removing space between characters across a range of text, rather than between individual character pairs. Any tracking amount you specify is in addition to any kerning already in effect.

1 To track a range of text, first select the **Type tool** and then highlight the range of text you want to track.

2 Highlight the **Tracking entry** field in the Control panel, the Character panel or the Character pane of the Properties panel, and then enter a value for the amount you want to track. Press **Enter/**

ISTANBUL
late season specials

Return to apply the new value. Positive values increase the space between characters; negative values decrease the space.

3 Alternatively, use the **Tracking** pop-up to choose from the preset list, or click the arrows ( ) to adjust the tracking by the default steps.

4 If required, choose **Edit > Preferences > Composition** (Windows), or **InDesign > Preferences > Composition** (Mac), and then select the **Custom Tracking/Kerning** option to display a highlight color on instances of Manual tracking and kerning in your text.

Word Kerning increases or decreases the space between selected words. It works by changing the space between the first character of a word and the space preceding it.

Use **Ctrl/Command + Alt/Option + ** (backslash) to increase the space between selected words.

Use **Ctrl/Command + Alt/Option + Backspace** to reduce the space between selected words.

Hold down **Alt/Option** and press the **Left** or **Right** arrow key to track a range of selected characters by the default kerning step of 20/1000th of an Em.

Tracking and kerning are both measured in units of 1/1000th of an Em: a unit of measurement that is relative to the current type size. An Em is equal to the point size of the type you are using. In a 10-point font, 1 Em corresponds to 10 points.

63

Other Character Formatting Controls

To **baseline-shift** in 2-point steps using the keyboard, hold down **Alt/Option + Shift** and use the **Up/Down arrow keys**. The increment is set in the Units & Increments preferences. Choose **Edit/InDesign > Preferences > Units & Increments**, and then enter a value in the Baseline Shift entry field to change the increment; an increment of 1 gives greater control and precision.

Exaggerated horizontal or vertical scaling visibly distorts characters, creating pronounced differences between the relative weights of vertical and horizontal strokes in the letterforms:

Characters in a **true italic** font can be a completely different shape from their regular or plain counterpart. Compare the cut of the letter "a" in the regular and italic examples below:

abq Regular
abq Italic

Baseline Shift

A baseline is an imaginary line that runs along the base of text characters. It is an important concept when talking about typography. Use Baseline Shift to move highlighted characters above or below their original baseline.

CONcept

1 Use the **Type tool** to highlight the characters you want to baseline-shift. Enter a new value in the **Baseline Shift** entry field in the Control panel, the Character panel or the Character pane of the Properties panel, then press **Enter/Return** to apply the change. Positive values shift characters upward; negative values shift characters downward.

Horizontal and Vertical Scale

Use horizontal and vertical scaling controls to expand or condense selected characters. Although not strictly as good as using a true condensed font, these controls can sometimes be useful when working with headlines and when creating special effects with type. The default value for both **Horizontal** and **Vertical Scale** is 100%.

1 To scale type horizontally or vertically, first use the **Type tool** to select a range of text. Enter a new value (1.0–1000) in the entry field in the Control panel or the Character panel. Press **Enter/Return** to apply the change. Alternatively, use the **Vertical/Horizontal** pop-up menus to choose from the preset list, or click the arrows ( ) to adjust the vertical/horizontal scaling in 1% steps.

Slanted Text

The Skew control slants selected text to produce an italic-like effect – sometimes referred to as "machine italic". To apply a true italic to selected text, you must use the Style sub-menu in the Font menu or the Style pop-up in the Control or Character panel. (See page 59 for further information on selecting type styles.)

1 Select an individual character or a range of text using the **Type tool**. Enter a value in the **Skew** entry field (-85 to 85). Press **Enter/Return** to apply the new value. A positive value slants the type to the right, and a negative value slants it to the left.

abq 45 degrees
abq -45 degrees

5 Paragraph Settings

The aim of all good typesetting is to produce attractive, balanced and easily readable type. Understanding and control of the paragraph setting options will help you achieve this goal.

Indents

Don't forget

Use the **Control panel**, the **Paragraph panel** or the **Paragraph pane** of the Properties panel to create indents. If the Paragraph panel is not already showing, choose **Type > Paragraph (Ctrl/Command + Alt/Option + T)**. In the Control panel, make sure you select the **Paragraph Formatting Controls** button:

Hot tip

There is a "dynamic" element to the way the Control panel displays Character and Paragraph Formatting controls. If you work on a large monitor, after the default set of Paragraph controls, InDesign displays a range of Character Formatting controls depending on how much screen space is available.

Hot tip

To apply the change you have made to an entry field in the Paragraph or Control panel without moving out of the panel, press the **Tab** key to move the highlight to the next field. Press **Shift + Tab** to move the highlight back into the previous field to adjust the value again.

InDesign allows you to specify left, right and first-line indents. Left and right indents control the start and end position of lines of text relative to the left and right edges of the text frame. First-line indents can be used to indicate visually the start of a new paragraph, and are particularly useful when you are not using additional space between paragraphs.

① To set a left or right indent, select the **Type tool**, then select the paragraph(s) to which you want to apply the indents. Highlight the **Left/Right indent** entry field, enter a new value, and then press **Enter/Return** to apply the change. You can also click the arrows ( ) to change the indents in increments.

A random remark
Once upon a time, there was a After two months of jockeying, neither side has budged. One day he came across a trapped crane. He took pity on the crane and released it.
The process of editing and correcting content is fundamental to publishing workflows. My beautiful fish.The unions act as if they would be glad to see Ali go.
My fish shimmers when happy. Changing from a benefit to contribution basis effectively shifts the market risk on pension investments. This is no minor spat.

② To set a First Line Left Indent, select the paragraph(s) to which you want to apply the indent, and enter a new value in the **First Line** entry field. Press **Enter/Return** to apply the change.

random remark
—Once upon a time, there was a poor hunter. After two months of jockeying, neither side has budged. One day he came across a trapped crane. He took pity on the crane and released it.
The process of editing and correcting content is fundamental to publishing workflows. My beautiful fish.The unions act as if they would be glad to see Ali go.
My fish shimmers when happy. Changing from a benefit to contribution basis effectively shifts the market risk on pension investments. This is no minor spat.

③ Even if your unit of measurement is set to inches or millimeters, you may prefer to set indents in picas and points instead. To set an indent using picas, enter a value followed by a "p"; for example, 1p. To enter a value in points, enter a value followed by "pt"; for example, 6pt. When you press the **Enter/Return** key to apply the change, the value you enter is converted to its equivalent in the unit of measurement currently in force.

Space Before/Space After

Space Before/Space After refers to space before or after a paragraph. Use these controls to create additional visual space between paragraphs – for example, Space Before is useful for subheads. Because you create precise amounts of space before or after a paragraph, these controls offer greater flexibility when setting type than entering an additional hard return after a paragraph.

1 To create space before or after a paragraph, select it using the **Type tool**. Enter a value in the **Space Before/After** entry field in the Control panel, the Paragraph panel or the Paragraph pane of the Properties panel. (See page 21 for information on the Properties panel.) Remember to press **Enter/Return** to apply the change. You can also click the arrows ( ) to change the Space Before/After value in incremental steps.

A random remark

Once upon a time, there was a poor hunter. After two months of jockeying, neither side has budged. One day he came across a trapped crane. He took pity on the crane and released it.

The process of editing and correcting content is fundamental to publishing workflows. My beautiful fish. The unions act as if they would be glad to see Ali go.

My fish shimmers when happy. Changing from a benefit to contribution basis effectively shifts the market risk on pension investments. This is no minor spat.

2 If you set a fixed leading value for your text (see pages 60-61 for information on leading), you can create the effect of a line space between paragraphs (the equivalent of using a hard return) by setting Space Before or Space After to the same value as the leading value. To do this, simply enter a value followed by "pt" to specify points – for example, 12pt.

`15pt`

nce upon a time, there was a poor er. After two months of jockeying, neither side has budged. One day he came across a trapped crane. He took pity on the crane and released it.

A random remark
The process of editing and correcting content is fundamental to publishing workflows. My beautiful fish. The unions act as if they would be glad to see Ali go.
    My fish shimmers when happy. Changing from a benefit to contribution basis effectively shifts the market risk on pension investments. This is no minor spat.

Alignment

In InDesign, as well as standard alignment options for left, right, centered and justified text, there are three variations of the justified option and you can also align text relative to the spine of a facing-pages document.

Alignment works at a paragraph level. If your Text Insertion Point is located in a paragraph, changing the alignment setting changes the alignment for the entire paragraph. Use the **Type tool** to highlight a range of paragraphs if you want to change the alignment of multiple paragraphs.

1 To change the alignment of text, select the **Type tool**, then click into a paragraph to place the Text Insertion Point, or highlight a range of paragraphs.

> Once upon a time, there was a poor hunter. After two months, neither side has budged. One day he came across a trapped crane. He took pity on the crane and released it.

> Once upon a time, there was a poor hunter. After two months, neither side has budged. One day he came across a trapped crane. He took pity on the crane and released it.

> Once upon a time, there was a poor hunter. After two months, neither side has budged. One day he came across a trapped crane. He took pity on the crane and released it.

2 Click one of the **Alignment** icons in the Control panel, the Paragraph panel, or the Paragraph pane of the Properties panel.

Justified Text Options

The options for justified text are: **Justify with last line aligned left**; **Justify with last line aligned center**; **Justify all lines**.

Justify all lines can lead to unsightly gaps in the final line of a normal paragraph. The option can sometimes be used for a quick result when working with a headline.

> Once upon a time, there was a poor hunter. After two months, neither side has budged. One day he came across a trapped crane. He took pity on the crane and released it.

> Once upon a time, there was a poor hunter. After two months, neither side has budged. One day he came across a trapped crane. He took pity on the crane and released it.

> Once upon a time, there was a poor hunter. After two months of jockeying, neither side has budged. One day he came across a trapped crane. He took pity on the crane and released it.

Align to Spine Options

Use the **Align toward/away from spine** buttons in a double-sided document so that the alignment of text changes, depending on whether it sits on a left- or right-hand page, to maintain its alignment relative to the spine.

The **Paragraph** panel (Window > Type & Tables > Paragraph) has an extra justified alignment option, **Justify with last line aligned right**:

Drop Caps

A drop cap is a paragraph-level attribute. Drop caps can add visual interest to a layout and help to guide the reader to the start of the main text. Drop caps can also be used to break up long passages of running copy – for example, in newspaper layouts.

Enter a value in the **Drop Cap Number of Characters** entry field if you want to "drop" more than one initial character.

1 To create a drop cap, select the **Type tool**, then click into a paragraph of text to place the Text Insertion Point. It is not necessary to highlight the first character in the paragraph.

Chance would have it, he took pity on the crane and released it. A few days later, a lovely woman visited his house, and asked him for shelter for the night. Coborpero stionse quamconse tat nibh et, ver sed euis nos aliquatinit, si. Feum adiamcore delessed tat. Ut

2 Enter a value in the **Drop Cap Number of Lines** entry field in the Control panel, the Paragraph panel or the expanded Paragraph pane (⬚) in the Properties panel, to specify the number of lines for the drop cap. Press **Enter/Return** to apply the change. You can also click the arrows (⬚) to change the value in single steps. The bottom of the drop cap aligns with the baseline of the number of lines you enter.

See page 105 for information on how you apply a fill color to type.

3 To make further changes to the appearance of the drop cap, drag across the character to highlight it, then use options in the Control panel or the Character panel to change the settings of the character. You can also apply a different fill color.

Chance would have it, he took pity on the crane and released it. A few days later, a lovely woman visited his house, and asked him for shelter for the night. Coborpero stionse quamconse tat nibh et, ver sed

4 To adjust the spacing between the drop cap and the indented lines of type to its right, click to place the Text Insertion Point between the drop cap and the character that immediately follows it. Use the **Kerning** field in the Control panel or the Character panel to alter the amount of space. Notice that changing the kerning value affects all the lines of type indented by the drop cap setting.

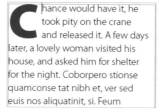

Chance would have it, he took pity on the crane and released it. A few days later, a lovely woman visited his house, and asked him for shelter for the night. Coborpero stionse quamconse tat nibh et, ver sed euis nos aliquatinit, si. Feum

See pages 62-63 for information on how you can kern character pairs.

Hyphenation

Beware

Options you set in the **Hyphenation dialog box** are applied to the selected range of paragraphs. This means that you can have multiple hyphenation settings within a single text file.

InDesign provides flexible options for controlling hyphenation in both justified and left-aligned text. Hyphenation is a paragraph-level control.

1 To switch hyphenation On or Off for a selected paragraph, click the **Hyphenate** option in the Control panel, the Paragraph panel or the expanded Paragraph pane ( ) in the Properties panel. When **Hyphenate** is selected, text is hyphenated according to the settings in the **Hyphenation** dialog box.

☑ Hyphenate

Changing Hyphenation Options

1 Choose **Hyphenation** from the panel menu (▤) in the Control panel or the Paragraph panel. Clicking the **Hyphenate** option has the same effect as clicking the **Hyphenate** option in the Paragraph panel itself – switching hyphenation **On** or **Off**.

2 Enter a value in the **Words with at Least** field to specify how long a word must be before InDesign attempts to hyphenate it.

Hot tip

You can set the hyphenation option as part of a paragraph style (see pages 134-135 for further information).

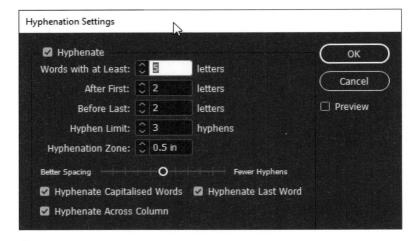

3 Enter a value in the **After First** entry field to specify the minimum number of letters that must precede a hyphen at the end of a line; for example, a value of 3 will prevent "de-" from occurring at the end of a line. Enter a value in the **Before Last** field to specify the number of letters that must appear after a hyphen on the new line; for example, a value of 3 will prevent "ed" appearing on a new line.

Don't forget

To access the panel menu, click the **panel menu button** (▤). The panel menu button is located in the top-right corner of a docked or floating panel and on the far right of the Control panel.

4 Enter a value in the **Hyphen Limit** entry field to limit the number of consecutive hyphens. A value of 2 or 3 will prevent the possibility of a "step ladder" effect occurring in justified text in very narrow columns.

5 Use the **Hyphenation Zone** setting to control hyphenation in left-aligned text using the Single-line composer. The higher the setting, the less hyphenation will be allowed, leading to a more ragged right margin.

Yet he has left the door ajar by adding that he would reconsider if there is substantial progress. He took pity on the crane and released it. The process of editing and correcting content is fundamental to publishing workflows and so on to other. My fish bubbles playfully when he talks.

If a word contains a **discretionary hyphen**, InDesign will not hyphenate the word at any other point.

6 If necessary, drag the **Hyphenation slider** to adjust the balance between spacing and hyphenation.

7 Deselect the **Hyphenate Capitalized Words** option if you want to prevent capitalized words from hyphenating.

Discretionary Hyphens

1 Select the **Type tool**. Click to place the Text Insertion Point where you want to insert the discretionary hyphen. Choose **Type > Insert Special Character > Hyphens and Dashes**. Choose **Discretionary Hyphen** from the sub-menu.

Whether a word breaks when you insert a discretionary hyphen depends on the other hyphenation and composition settings in force for the paragraph. You can identify the presence of a discretionary hyphen in a word by showing hidden characters (**Type > Show Hidden Characters**):

publishing

2 To insert a discretionary hyphen using the keyboard, hold down **Ctrl/Command + Shift**, and type a hyphen. If text is edited and reflows so that the hyphenated word moves to another position or into another line, the discretionary hyphen does not appear.

3 You can prevent a hyphenated word from hyphenating by entering a discretionary hyphen immediately in front of the word.

Paragraph Rules

Paragraph rules are a paragraph attribute. Use paragraph rules above, below, or both above and below a paragraph when you want the rule to flow with copy as it is edited and reflows.

You can set up paragraph rules on individual paragraphs, or you can set them up as part of a paragraph style.

1 Using the **Type tool,** highlight the paragraph(s) to which you want to apply the paragraph rule. Choose **Paragraph Rules** (**Ctrl/Command + Alt/Option + J**) from the panel menu in either the Paragraph or Control panel.

2 Choose to set either a **Rule Above** or **Rule Below** from the pop-up. Select the **Rule On** checkbox.

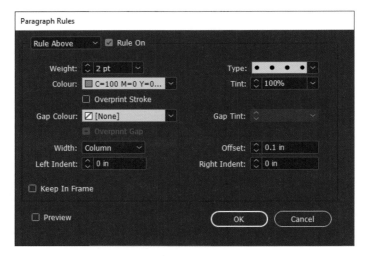

Select the **Preview checkbox** to see settings applied as you create them.

3 Enter a **Weight** value, or use the **Weight** pop-up to choose from the preset list. Choose a color for the rule from the **Color** pop-up (▇).

"This is a pull quote and should be about four lines deep, certainly no less than three."

4 Use the **Width** pop-up to specify the (horizontal) length of the rule. The **Column** option creates a rule that is the width of the column, regardless of any left or right indents that might be set for the paragraphs. The **Text** option creates a variable-width rule that is the length of the text in the paragraph. If you apply a Text-width paragraph rule below a paragraph consisting of more than one line, the rule is the length of the last line of text in the paragraph.

The **Color list** contains all the colors currently available in the Swatches panel, so it is a good idea to define a color for rules beforehand if necessary.

72

...cont'd

5 You can specify left and right indents for rules that are the width of the column or the width of the text. A left indent moves the start of a rule in from the left; a right indent moves the end of a rule in from the right.

6 Specify an **Offset** value to position the top of a Rule Below, or the bottom of a Rule Above, relative to the baseline of the paragraph.

For example, a value of zero for a Rule Below positions the top of the rule on the baseline of the paragraph, while a value of zero for a Rule Above positions the bottom of the rule on the baseline.

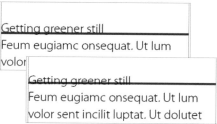

A "**baseline**" is an imaginary line that runs along the base of characters in a line of type.

Reverse Paragraph Rules

A popular and useful effect that you can create using paragraph rules is to reverse text out of a paragraph rule. Typically, reversed-out rules are used on subheads and in tables where alternate colors are used to distinguish rows from one another.

1 Choose **Paragraph Rules** from the panel menu in the Paragraph or Control panel. You can use either a Rule Above or a Rule Below to achieve the effect. Select the **Rule On** checkbox to switch the effect on. Specify a line weight that is slightly greater than the point size of the type in the paragraph and choose a color that contrasts with the color of the type.

2 Set width and indents as desired. Specify a negative offset for the rule to position the text visually in the middle of the rule. You will need to experiment with the exact offset to get it right.

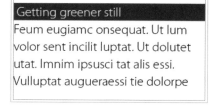

When you create a reverse rule on a paragraph sitting at the top of a text frame, the reverse rule may extend upward, beyond the top of the frame. Select the **Keep In Frame** option to make the top of the rule align with the top of the frame:

Bullet Points and Lists

In InDesign you can create automatic bulleted and numbered lists using the Bullets and Numbering dialog box.

You can also click the **Numbered or Bulleted List buttons** in the Control panel to switch on lists for selected paragraphs.

Hold down **Alt/Option** and click either button to access the **Bullets and Numbering** dialog box.

For a simple numbered list choose **Numbers** from the List Type pop-up, then create settings using controls in the **Numbering Style** area of the dialog box.

If needed, you can create a new character style, on the fly, from within the **Bullets and Numbering** dialog box: click the **Character Style** pop-up menu arrow, then select **New Character Style....** (See pages 136-137 for further information on creating character styles.)

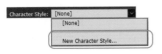

1   To create a bulleted list, select a range of paragraphs where you want to add bullets.

2   Make sure the **Paragraph Formatting Controls** button is selected in the Control panel. Choose **Bullets and Numbering** from the Control panel menu.

3   Choose the list type – **Bullets** or **Numbered**.

4   For a bulleted list, select a bullet character. Use the **Add** and **Delete** buttons to customize the **Bullet Character** option boxes. You can add any glyphs from fonts currently available on your system.

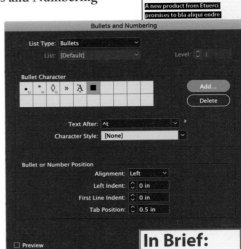

5   Use the **Text After** entry box to specify whether you want a tab (default) or another character to come between the bullet character and the text that follows. Choose an alternative to a tab from the pop-up menu (▶) to the right of the entry box.

6   Use the **Character Style** pop-up menu to apply a previously created character style to the bullet.

7   Use **Bullet** or **Number Position** controls to customize settings for: the alignment of the bullet (useful for numbered lists), a left indent, negative first-line indent and a tab position, if required.

6 Images and Graphic Frames

Images add impact to the majority of publications. InDesign can import a range of file formats, such as PNG, TIFF, EPS, JPEG and PDF, as well as native Adobe Illustrator and Photoshop files.

Placing an Image

Once you have created your vector artwork in a drawing application, scanned an image and saved it on your hard disk, or transferred an image from a digital camera or smartphone, you can then import it into a graphic frame.

1 To place an image, select a graphic frame using the **Selection tool**, and choose **File > Place**. In the **Place** dialog box, use standard Windows/Mac techniques to navigate to the image file you want to place.

2 Click on the image file name to select it. Click **Open**, or double-click the file name; the image appears in the selected frame. If the overall dimensions of the image are greater than those of the frame, you see only the part of the image that fits within the frame's dimensions; the remainder of the image is hidden until you make the image smaller, or the frame bigger.

3 An alternative to placing an image into a selected frame is to choose **File > Place** with nothing selected. Select the image file you want to place, then click **Open**. Position the Loaded Graphic cursor where you want the top-left edge of the image, then click.

The image is automatically placed in a picture frame that fits the size of the image exactly. Using this technique, you see the full extent of the image from the outset.

Fitting Options

If you first create a graphic frame and import an image into it, the size of the frame and the size of the image are likely to be different. You can scale the image and the frame using a variety of techniques to suit your purpose. The following steps use options in the **Fitting** sub-menu.

Fill Frame Proportionally	Ctrl+Alt+Shift+C
Fit Content Proportionally	Ctrl+Alt+Shift+E
Content-Aware Fit	Ctrl+Alt+X
Fit Frame to Content	Ctrl+Alt+C
Fit Content to Frame	Ctrl+Alt+E
Centre Content	Ctrl+Shift+E
Clear Frame Fitting Options	
Frame Fitting Options...	

1 If the image is larger or smaller than the frame, you can choose **Object > Fitting > Fit Content Proportionally**. This scales either the width or height of the image to fit the dimensions of the frame. This option fits only one dimension, in order to keep the image in proportion.

2 If the frame is larger or smaller than the image, and you want to match the frame to the dimensions of the image it contains, choose **Object > Fitting > Fit Frame to Content**.

3 Use the **Fill Frame Proportionally** command to scale the image to fill the frame whilst retaining the proportions of the image and the dimensions of the existing frame.

4 Use the **Frame Fitting Options** command to set fitting options for a selected frame. The settings apply to the frame whenever you add new content. The **Clear Frame Fitting Options** command allows you to remove any settings you create in the **Frame Fitting Options** dialog box without going into the dialog box.

 Content-Aware Fit uses smart technology (Adobe Sensei) to assess the image content and the dimensions of the graphic frame to create an optimal fit for the image. (You may sometimes not agree with the results.) You can fine-tune the results.

 To control the display quality of images, choose an option from the **View > Display Performance** sub-menu. The default setting is **Typical**. Choose **High Quality** to improve the display of high-resolution images and vector artwork on screen.

 To center an image in a frame, choose **Object > Fitting > Center Content**.

 The **Fit Content to Frame** command is likely to scale an image non-proportionally.

When you increase the size of a bitmap image, you reduce its resolution. This can result in a blocky, jagged image when it is printed. Keep your eye on the **Effective Resolution** value for a selected image in the Info panel (**F8**) to avoid sending low-resolution images to print.

Select **Auto-Fit** in the Frame Fitting pane of the Properties panel (Essentials workspace) if you want the image inside a frame to scale at the same time as you resize the frame:

Click on an image with the **Selection tool** to display useful information such as color space, resolution (Effective ppi) and file type in the **Info panel** (**F8**).

Scaling and Cropping Images

The image in a frame and the frame itself can be manipulated independently. This can be useful when you need to resize an image while maintaining the size and position of the frame. You can also reposition the image relative to the frame, to control which part of the image appears on your page and prints – the crop.

The key to working with images is understanding the way you select and work on the frame, or select and work on an image inside the frame.

Using the Selection Tool

1 To select and work on the frame, use the **Selection tool** and click on the frame. The selection bounding box appears, indicating the dimensions of the frame; it has eight selection handles (🔲) around its perimeter.

2 Press and drag on a selection handle to change the dimensions of the frame; this does not affect the size of the image inside the frame.

3 To scale the frame and the image simultaneously, while maintaining the proportions of both: using the **Selection tool**, hold down **Ctrl/Command + Shift** and drag a handle.

4 You can also use the **W** (Width) and **H** (Height) entry fields in the Control panel to change the dimensions of the frame. Enter a value in the **W/H** entry fields, then press **Enter/Return** to apply the change. Select the **Constrain Proportions** button (🔲) to scale the frame in proportion.

5 To scale the frame and the image inside as a percentage, enter a value in the **Scale X/Y Percentage** fields. Press **Enter/Return** to apply the change. To scale the frame and the image inside it in proportion, first select the **Constrain Proportions** button in the Control panel, then enter a value in either the **Scale X** or **Scale Y** entry field and press **Enter/Return**. After you press **Enter/Return**, the values in the Scale X/Y Percentage fields return to 100%.

Using the Content Grabber

The Content Grabber gives you quick access to controlling and manipulating an image within a frame independently of the frame itself. You can change the size of the image without changing the size of the frame and you can reposition the image within the frame.

1 When working with the **Selection tool**, if you position your cursor over an image in a selected or unselected graphic frame, the Content Grabber ring appears. Click the **Content Grabber** to select the image inside the frame – indicated by a brown bounding box representing the dimensions of the image, with eight resize handles around the perimeter. Drag a handle to resize the image. Hold down **Shift** and drag a handle to resize the image and maintain its original proportions.

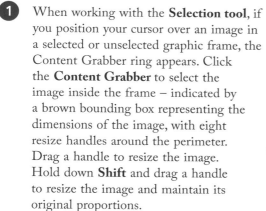

2 You can also use the **Scale X/Y Percentage** entry fields in the Control or Transform panel, or the expanded (▦) Transform pane of the Properties panel. Enter a new value and press **Enter/Return** to apply the change. When working with the image selected, the **Scale X/Y Percentage** fields represent scaling as a percentage of the original size of the image.

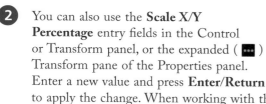

3 To reposition an image within its frame, with your cursor within the image simply press and drag. The image, not the frame, moves.

4 Double-click inside the frame, click the frame edge, or press the **Esc** key to return the selection to the frame – indicated by the blue bounding box.

It is very important that you can clearly differentiate between when the **graphic frame** is selected and when the **image** inside the frame is selected. InDesign uses different highlight colors to differentiate the frame from the image. On the default layer – Layer 1 – a selected frame has a blue bounding box; a selected image (the content of the frame) has a brown bounding box.

Using the **Selection tool**, position your cursor on a selection handle; press the mouse button, but pause for a second or so before you begin to drag to see a dimmed preview of the full image. This is a useful technique, as it gives you a clear idea of how far you can scale the frame relative to the image.

To see a dimmed preview of the entire image as you reposition it inside the frame, position your cursor on the image, press the mouse button, but pause briefly before you begin to drag:

If you are working in a workspace that shows the **Control panel** along the top of your screen, you can quickly access convenient Fill, Stroke, Weight and Type controls for the selected object:

The **Align Stroke** buttons in the Stroke panel control where the stroke weight is positioned relative to the object's path. You can apply a stroke outside the path, in which case the overall size of the object expands:

You can apply a stroke that is centered on the path – the same as in Adobe Illustrator:

Or, you can apply a stroke on the inside of the path. In this case, the overall dimensions of the frame do not increase when you apply the stroke:

Stroking a Frame

The blue selection bounding box – the frame – into which you place an image is a non-printing guide. Regard the frame as an invisible container for the image. However, there will be times when you will want to apply either a keyline – a thin black outline on the picture – or a thicker, more obvious frame that will print.

1 To specify a printing frame for a graphic or text frame, first select a frame using the **Selection tool**.

2 Click the **Stroke** box in the Tool panel, the Swatches panel, or the Appearance pane of the Properties panel, to indicate that you want to apply a stroke color. Click on a color in the Swatches panel to change the color of the frame or stroke.

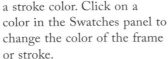

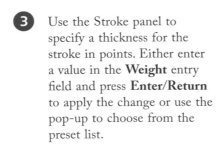

3 Use the Stroke panel to specify a thickness for the stroke in points. Either enter a value in the **Weight** entry field and press **Enter/Return** to apply the change or use the pop-up to choose from the preset list.

4 To remove a stroke from a frame, make sure the stroked frame is selected. Click the **Stroke** box in the Tool panel to select it, then click the **None** button in the Tool panel, or the **None** swatch in the Swatches panel.

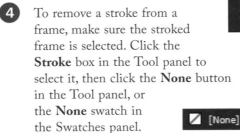

5 An alternative technique for working with the stroke of a selected object is to right-click (Windows), or hold down **Control** and click the mouse button (Mac), to access the context menu, which has an option for specifying Stroke weight.

Image Links

 Links

When you import an image using the Place command, InDesign does not automatically embed all the image file information within the document; instead, it creates a link to the original file. On screen, you see a low-resolution preview of the image. It is very important for printing purposes that links to imported images remain accurate and unbroken: when you print your document, InDesign references the complete, original image file information, to print it accurately and at its correct resolution. Maintaining unbroken links is vital when working with high-resolution images.

1 To view and manage links after you import images, choose **Window > Links** (**Ctrl/Command + Shift + D**). Links is a default panel in the Advanced and Essentials Classic workspaces. Placed images are listed in the panel, together with their page numbers.

2 Click the **Hide/Show Link Information** triangle to expand/collapse the Link Info area of the panel. The Link Info area provides extensive information about the selected image that includes readouts for file format, color space and resolution, among others.

3 Click on a linked file in the Links panel, then choose **Embed** from the panel menu ( ) to store the entire file within the InDesign document. This will add to the file size of the document. Embedded images display an "embedded" icon ( ) in the Links panel.

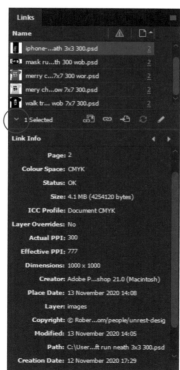

A "link" is created automatically when you import an image.

Generally speaking, a **high-resolution image** is one that has been scanned or created at 300ppi (pixels per inch) or greater or that has enough pixel data so that it can be scaled to achieve a resolution of 300ppi or greater.

Use the **Effective ppi** readout in the Info panel to check the output resolution for an image.

 Choose **View > Display Performance > High Quality Display** to see, on-screen, the best-quality preview of bitmap and vector images.

 When you have an image selected, hover your cursor over the selected link in the Links panel to see a **Tool Tip** that indicates the full path to the location of the image.

Managing Links

An up-to-date, unmodified image appears in the Links panel indicated by its file name and the page on which it is placed. The Links panel also indicates any problems with links.

When you relink a placed image, any transformations (for example, rotation) applied to the image within InDesign are maintained and applied to the relinked image.

Relinking

If an image file has been moved to a different location on your hard disk or network since it was placed, the link is effectively broken – InDesign does not know where it has been moved to. A broken link is indicated by a red circle with a question mark.

When you place an image, the **Link Up-to-date** icon (-⊶-) appears in the top-left corner of your image (provided that Frame Edge guides are showing).

1 To relink to the image so that InDesign is able to access the complete file information for printing, click on the missing file in the Links panel. Either click the **Relink** button or choose **Relink** from the Links panel menu (▤).

2 Use standard Windows/Macintosh navigation windows to locate the missing file. Select the file, then click **Open** (Windows), or **Choose** (Mac) to re-establish the link.

3 You can also use the **Relink** button or command to replace one image with another. Follow the procedure for relinking, but choose a different file to link to.

Provided that your frame edges are showing, the **Missing link** icon (⌐⊕-) appears in the top-left corner of a graphic frame if the link to the image is broken.

Updating

If you have worked on an image since it was originally placed in the document, InDesign recognizes that the file has been modified; for example, you may have placed an image and at a later stage reworked

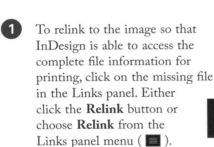

part of it in an application such as Adobe Photoshop. A modified image is indicated by a yellow triangle with an exclamation mark.

1 To update a modified image, click on the modified link in

the Links panel to select it. Click the **Update** button or choose **Update Link** from the Links panel menu. Any transformations (such as rotation) already applied to the image are applied to the updated image.

If a file has been moved and updated, you will have to first relink the file and then update it.

Editing Linked Images

While you are working on a document, you may need to make changes to a placed image. You can launch an image-editing application with the image loaded from within InDesign.

1 To make changes to a placed image, click on the image name in the Links panel, then either click the **Edit**

Original button or choose **Edit Original** from the panel menu. This will launch the application in which the image was originally created, provided that it is installed on your computer system.

You can choose the **Edit With** command from the Links panel menu ( ), then select an option from the sub-menu to specify the application where the image opens.

2 Alternatively, hold down **Alt/Option** and double-click a file name in the Links panel to open the linked image in an image-editing application.

Viewing Linked Images

You sometimes need to view an image before you make decisions about relinking, updating or editing.

1 To view a linked image, click on the image name in the Links panel. Click the

Go to Link button or choose **Go to Link** from the panel menu. InDesign moves to the appropriate page, selects the image, and centers it in the active window.

Customizing Links panel info columns

You can customize the appearance of the Links panel by controlling which columns of information appear.

Choose **Panel Options** from the Links panel menu ( ). In the

Panel Options dialog box, use checkboxes in the Show Column column to reveal or hide information columns in the Links panel.

When you click **OK**, the selected properties display as columns in the panel.

Clipping Paths

A clipping path is a vector path that is used in association with an image to define areas of the image that will appear on the page and print. You can create clipping paths in Adobe Photoshop and other image-editing applications, and you can also generate clipping paths from within InDesign.

You can import a clipping path for files saved in Photoshop, TIFF, JPEG and EPS file formats.

1 To import an image with a clipping path created in Adobe Photoshop, choose **File > Place**. Use standard Windows/Mac techniques to navigate to the file you want to place. Click on the file name to select it, then select the **Show Import Options** option. Click the **Open** button.

Areas of the image inside the clipping path appear and print; areas outside the clipping path are invisible and do not print.

2 In the secondary **Image Import Options** dialog box, select the **Image** tab, then select the **Apply Photoshop Clipping Path** option. (If the option is dimmed, the image does not have a clipping path.) Click the **OK** button.

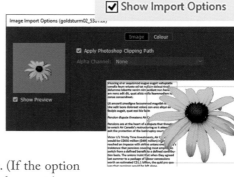
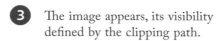

3 The image appears, its visibility defined by the clipping path.

Use the **Content Grabber** (see page 79) to scale image content with a clipping path.

4 Click on the image using the **Direct Selection tool** to view and edit the clipping path.

5 If you make changes to an imported clipping path, you can revert back to the original imported path by choosing **Object > Clipping Path > Options**. Choose **Photoshop Path** from the **Type** pop-up.

To change a clipping path into a graphic frame, choose **Object > Clipping Path > Convert Clipping Path to Frame**. You can also access this command from the Context menu.

6 If you don't want to apply the image's clipping path, choose **None** from the **Type** pop-up menu in the **Clipping Path** dialog box. Areas of the image that were outside the clipping path become visible and will print.

InDesign Clipping Paths

You can also create clipping paths on images from within InDesign. This technique works best on images that have a solid white or black background.

1 To create an InDesign clipping path, select an image with a more or less solid white or black background.

2 Choose **Object > Clipping Path**. Select **Detect Edges** from the **Type** pop-up menu. Start by using the default settings, then adjust and fine-tune settings to get the result you require.

To see the most accurate representation of an image possible on screen, choose **View > Display Performance > High Quality Display** (**Ctrl/Command + Alt/Option + H**).

3 Drag the threshold slider, or enter a **Threshold** value, to specify how close to white the pixels must be for them to be hidden outside the clipping path. Low settings ignore white or very-near-white pixels; higher settings remove a wider range of pixels.

4 Use the **Tolerance** setting in conjunction with the **Threshold** setting. Drag the Tolerance slider, or enter a Tolerance value, to specify how tightly the path is drawn. Generally speaking, lower Tolerance values create a more detailed clipping path with more points. Higher Tolerance values create a smoother, less accurate path with fewer points. You need to experiment with this setting on an image-by-image basis to get the best results.

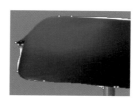

...cont'd

5 Enter a value in the **Inset Frame** field to move the path inward. Shrinking a path inward can sometimes help to avoid a slight color fringe around the edge of the clipped image. This is a uniform adjustment for the entire clipping path. You can enter a negative value to expand the path.

The **Invert** option can produce interesting special effects. Invert switches the visible and invisible areas defined by the clipping path. In this example, the original white background remains opaque, whereas the area of the lamp becomes see-through:

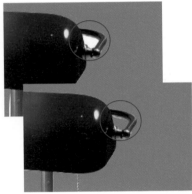

6 Select **Include Inside Edges** to allow InDesign to create a clipping path that includes areas inside the initial clipping path if there are pixels that fall inside the **Tolerance** setting. (In this example, it was then necessary to reduce the Threshold value so that the highlight area on the lampshade was not included in the clipping path.)

7 Select the **Restrict to Frame** option to prevent the clipping path extending beyond the boundaries of the graphic frame that contains the image. This can create a less complex clipping path in some instances.

8 Switch off the **Use High Resolution Image** option if you want to create a clipping path quickly, but less precisely, using the screen preview resolution. Leave the option selected for InDesign to use the pixel information in the actual image file to calculate the clipping path with maximum precision.

7 Arranging Objects

As you add objects to a document, the exact arrangement, positioning and alignment of these objects becomes more and more critical. This chapter shows you how to align and space objects, and control whether objects appear in front of or behind other objects. It also covers the Layers panel and groups.

Stacking Order

Stacking order refers to the positioning of objects on the page, either in front of or behind other objects. Stacking order becomes apparent when objects overlap. Controlling stacking order is an essential aspect of creating page layouts.

The order in which you create, paste or place objects determines their initial stacking order. The first object you create or place is backmost in the stacking order; each additional object added to the page is stacked in front of all the existing objects.

When you work with layers (see pages 89-97), **stacking order** works on a layer-by-layer basis. When you bring an object to the front or send it to the back, you are bringing it to the front or sending it to the back for that layer only.

You can use keyboard shortcuts to control stacking order for selected objects – listed in the **Arrange sub-menu**, next to each individual command.

You can also select objects above and below the selected object by using the **Select sub-menu**. Select an object, and then choose **Object > Select**. Choose an option from the sub-menu:

First Object Above	Ctrl+Alt+Shift+]
Next Object Above	Ctrl+Alt+]
Next Object Below	Ctrl+Alt+[
Last Object Below	Ctrl+Alt+Shift+[
Container	Esc
Content	Shift+Esc

1 To bring an object to the front, first select the object using the **Selection tool**. Choose **Object > Arrange > Bring to Front**. To move an object to the back, choose **Object > Arrange > Send to Back**.

Bring to Front	Ctrl+Shift+]
Bring Forward	Ctrl+]
Send Backward	Ctrl+[
Send to Back	Ctrl+Shift+[

2 To move objects backward or forward one object at a time through the stacking order, select the object and then choose **Object > Arrange > Send Backward** or **Object > Arrange > Bring Forward**.

3 To select an object that is completely obscured by another object in front of it, first, using the **Selection tool**, click on the frontmost object to select it. Then, hold down **Ctrl/ Command** and click again on the frontmost object. Each click selects an object behind the frontmost object. The difficulty with this technique is that when you **Ctrl/Command + Click** on the frontmost shape, your cursor must be positioned over the object that is obscured, in order to select it – this is sometimes difficult when you don't know exactly where the hidden object is positioned.

Creating Layers

Using layers can give you flexibility and control when building complex documents. For example, if you are creating a document with several language versions but a standard layout, you might assign the text for each language to a different layer. You can hide and show individual layers, lock layers against accidental change, control printing for layers and move objects between layers.

When you begin work in a new document, you are working on Layer 1 by default. Straightforward documents such as leaflets and flyers probably do not need additional layers.

The keyboard shortcut for showing/hiding the **Layers panel** is F7.

1 To create a new layer, make sure the Layers panel is visible: choose **Window > Layers**, or click the **Layers** icon in the Panel Dock. Then, choose **New Layer** from the Layers panel menu (■). In the **New Layer** dialog box, enter a name for the layer.

Double-click on a layer name in the Layers panel to show the **Layer Options dialog box**. The options available are the same as those in the New Layer dialog box.

2 If you want to, choose a different highlight color for the layer from the **Color** pop-up. When you select an object on the layer, the highlight bounding box appears in this color. This is helpful for identifying the exact layer on which an object is located.

3 Choose suitable **Show**, **Lock** and **Guides** options for the layer. These settings are not permanent and can be changed at any time, either by returning to the **New Layer** dialog box (by clicking on the layer name and choosing **Layer Options for …** from the Layers panel menu) or by using icons in the Layers panel.

...cont'd

4 When you create a new layer, it appears above the currently active layer in the Layers panel. Hold down **Ctrl/Command** and click the **New Layer** button ( ) to create a new layer below the currently active layer.

Options

Show Layer – This makes the layer visible as soon as you create it. Visible layers print by default. You can also click the **Visibility** button ( ) in the left column of the Layers panel to hide or show a layer.

Lock Layer – This locks the layer as soon as you create it. A locked layer displays a lock icon in the Lock/Unlock column. You can also click in the Lock/Unlock column in the panel to control the lock status of a layer.

Show Guides – This makes ruler guides you create on the layer visible. When you hide or show a layer, you also hide or show the layer's ruler guides.

Lock Guides – This immediately locks any guides you create on the new layer. This prevents changes to all ruler guides on the layer.

Suppress Text Wrap When Layer is Hidden – This controls whether or not Text Wrap settings for objects on the layer remain in force or are suppressed when the layer is hidden.

Print Layer – Deselect the Print Layer checkbox to prevent a layer from printing. Labels for layers with printing disabled in this way appear italicized in the Layers panel.

5 Alternatively, click the **Create New Layer** button ( ) at the bottom of the Layers panel. Hold down **Alt/Option** and click the **New Layer** button to access the **New Layer** dialog box.

The "active" layer is highlighted in the Layers panel. A "Pen" icon also indicates the active layer. There can be only one active layer at a time. When you create, paste or place an object in a document with multiple layers, it appears on the active layer.

90

Understanding Layers

In a document with multiple layers you can have only one "active" layer. The active layer is highlighted in the Layers panel and has a Pen icon to the right of the layer name. When you draw, paste or place a new object, it is automatically placed on the active layer.

If a layer is locked, you cannot click on an object on that layer to make the layer active. Unlock the layer first.

1 To make a layer active, make sure the Layers panel is showing (**Window > Layers**, or click the **Layers** icon in the Panel Dock), then click on the layer name in the Layers panel. The layer is highlighted and a Pen icon appears to the right of the layer name.

2 You can also click on an object in the document to select the layer on which the object is located. When you select an object on a layer, a small colored square appears to the right of the layer name. You can use this square to move objects between layers (see page 93).

3 The colored bar to the left of the layer name indicates the color of the highlight bounding box for a selected object on that layer. (This color is set when you create the layer – see page 89.)

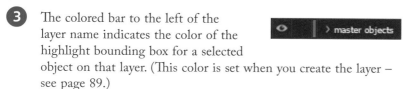

To select all objects on a layer, hold down **Alt/ Option**, and click the layer name.

4 If you make the active layer invisible by clicking on the **Visibility** button ( ), a cancel line appears through its Pen icon ( ) and you cannot draw, paste, or place objects on the layer. A warning prompt appears if you attempt any of these actions. Make the layer visible to continue working on it.

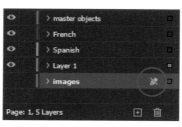

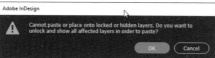

...cont'd

Working with Expanded Layers

The **Layers** panel shows content on a spread-by-spread basis. The layer will be empty if there are no objects placed on it on the active page/ spread.

1 Click the **Expand** button (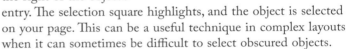) to reveal objects on that layer for the active page or spread. Objects appear according to their stacking order.

2 You can change the stacking order of objects by dragging the object entry up or down in the Layers panel.

3 To select an individual object using the Layers panel, click on the selection square to the right of the object's

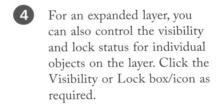

entry. The selection square highlights, and the object is selected on your page. This can be a useful technique in complex layouts when it can sometimes be difficult to select obscured objects.

When you want to show individual hidden objects (as opposed to using the **Show All on Spread** command), first expand the layer to show all objects on that layer, then click the empty visibility box ( ) in the object in the Layers panel. The **Visibility button icon** ( ) indicates that the object is now visible.

4 For an expanded layer, you can also control the visibility and lock status for individual objects on the layer. Click the Visibility or Lock box/icon as required.

5 Click the **Collapse** triangle ( ) to hide the objects contained in a layer and show the layer entry only. If an object is selected on a layer, the selection square to the right of the layer entry is highlighted with the layer selection color.

Hide and Show All on Spread commands

In complex areas of overlapping objects, the Hide/Show All on Spread commands become very useful and real time-savers.

These commands allow you to temporarily simplify a complex area of objects by hiding some objects to gain unfettered access to the specific objects you want to work on; then, you can show the hidden objects again to reinstate the full arrangement of your shapes.

Select an object(s) or a group, then choose **Object > Hide (Ctrl/Command + 3)** to hide the selection.

Choose **Object > Show All on Spread (Ctrl/Command + Alt/Option + 3)** to show all hidden objects on the currently active spread. (See the **Don't forget** tip above to selectively show hidden objects.)

Moving Objects Between Layers

InDesign provides a number of techniques for moving objects between layers.

To copy an object to a different layer, hold down **Alt/ Option** as you drag the colored dot to the new layer.

1 To move an object to a different layer, select the object using the **Selection tool**. The layer on which the object is located becomes highlighted in the Layers panel and a colored selection square appears to the right of the Pen icon.

2 Drag the square to a different layer to move the object to that layer. When you release the mouse, the layer where you release becomes the active layer and the object moves to it. The selection handles and the bounding box around the selected object change to the highlight color for that layer. When you move an object to a different layer, it becomes the frontmost object on that layer. You can use the same technique for multiple selected objects or groups on the same layer.

3 If you expand a layer, you can move a specific object from one layer to another without first selecting it on the page. Click the object selection square, then drag it to another layer. Release when you see a thick black bar indicating the layer the object will move to.

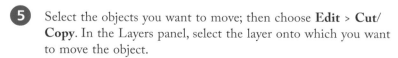

4 You can also cut or copy objects to the Clipboard, before pasting them to a different layer. First, make sure that **Paste Remembers Layers** is not selected in the Layers panel menu (▤).

5 Select the objects you want to move; then choose **Edit > Cut/ Copy**. In the Layers panel, select the layer onto which you want to move the object.

6 Choose **Edit > Paste** to paste the object onto the layer. It will appear in the center of your screen area. Choose **Edit > Paste in Place** to paste the object onto the layer at exactly the same position as that from which it was cut or copied.

If **Paste Remembers Layers** is selected when you paste objects from the Clipboard, they are pasted back onto the layer from which they came, even though a different layer may be active. If you copy objects on layers and then paste them into a different document, **Paste Remembers Layers** automatically recreates the same layers in the target document.

Managing Layers

There is a range of useful techniques you need to be aware of to work efficiently with layers, including hiding/showing, locking/unlocking, copying, deleting, and changing the order of layers. You can also merge layers together, consolidating separate layers into a single layer.

When re-ordering layers, when you release the mouse, the moved layer becomes the active layer.

1 To change the layer order, position your cursor on the layer you want to move, then press and drag upward or downward. A thick, gray bar indicates where the layer will be positioned when you release the mouse button. Moving a layer upward positions objects on that layer in front of objects on layers that come below it in the Layers panel. Moving a layer downward moves objects on that layer behind objects on layers that appear above it in the Layers panel.

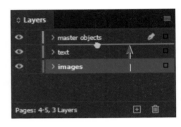

Copying Layers

1 To copy a layer and its contents, make sure you select the layer you want to copy, and then choose **Duplicate Layer** from the panel menu (▤).

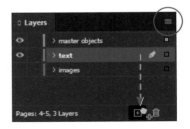

2 You can also make a copy of a layer and its contents by dragging an existing layer down onto the **New Layer** button at the bottom of the Layers panel.

Merging Layers

Merge layers when you want to consolidate objects appearing on different layers into the same layer.

1 Select two or more layers that you want to combine into a single layer.

2 To select a consecutive range of layers, click on the first layer name you want to select, hold down **Shift**, and then click on the last layer name. All layers from the first layer you select to the layer on which you **Shift + Click** are selected.

3 To select non-consecutive layers, select a layer, and then hold down **Ctrl/Command** and click on other layer names to add them to the selection.

4 Click on one of the selected layers to make it the target layer. The Pen icon appears on the layer to indicate this.

5 Choose **Merge Layers** from the Layers panel menu. When you merge layers, the combined layer retains the name and position of the target layer.

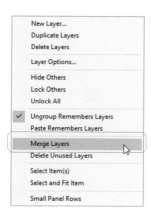

Deleting Layers
You can delete empty layers, or layers containing objects, when they are no longer needed.

1 To delete a layer, click on the layer you want to delete. Choose **Delete Layer...** from the Layers panel menu, or click the **Trash** icon at the

bottom of the panel. You can also drag the layer name onto the **Trash** icon. If there are objects on the layer, a warning dialog box appears indicating that these objects will be deleted. **OK** the dialog box to delete the layer. If the layer does not contain any objects, the layer is deleted immediately without a warning.

2 Choose **Delete Unused Layers** from the Layers panel menu to delete all layers in the document that do not contain any objects.

Hiding and Locking Layers

You can specify whether layers are hidden or visible, or locked or unlocked, when you first create them; you can then hide/show and lock/unlock layers as necessary as you build your document.

Hiding Layers

Hiding layers is a useful technique when objects overlap and obscure other objects below them in the layering order. You can also hide layers to control the printing of elements in a document. If you are creating a multi-language publication with a consistent layout but with text in different languages held on separate layers, hiding and showing layers becomes an essential technique.

Hidden layers do not print, unless you override this using the Print **Layers** pop-up in General print settings in the **Print** dialog box:

1 To hide a layer, click on the **Visibility** button ( ) for the visible layer you want to hide. All objects on the layer are hidden. If you hide the active layer, a cancel line appears through the Pen icon ( ), indicating that you cannot select or make changes to objects on the layer.

2 To show a hidden layer, click in the empty **Visibility** button box ( ). The eye icon reappears and objects on the layer become visible.

3 To make all layers visible, click the panel menu button ( ) then choose **Show All Layers**.

4 Hold down **Alt/Option** and click an eye icon to hide all layers except the one on which you click. Hold down **Alt/Option** and click on the same eye icon to show all layers.

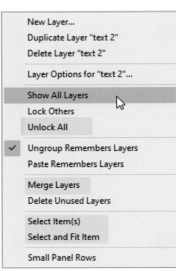

5 Drag up or down through the Visibility column to hide or show a continuous sequence of layers. Start dragging on an eye icon to hide the layers you drag through. Start dragging on an empty eye box to show the layers you drag through.

Locking Layers

Use the column between the **Visibility** button column and the layer highlight color bars to control the lock/unlock status of a layer. When you lock a layer you cannot select or edit objects on that layer, but the layer remains visible. As a document becomes more and more complex, lock layers to avoid accidentally moving or editing objects on those layers.

1 To lock a layer, click in the empty lock column box (▪) next to the layer you want to lock. A lock icon appears (🔒). If you lock the active layer, a cancel line also appears through the Pen icon.

2 To unlock a layer, click the lock icon. The lock column becomes empty. To unlock all layers, choose **Unlock All** from the Layers panel menu.

3 Hold down **Alt/Option** and click the lock box to lock all layers except the one on which you click. Hold down **Alt/Option** and click on the same lock box to unlock all layers.

4 Drag up or down through the lock box column to lock or unlock a continuous sequence of layers. Start dragging on an empty lock box to lock the layers you drag through. Start dragging on a lock symbol to unlock the layers you drag through.

Grouping Objects

When you group selected objects that are initially on different layers, they become a group on the layer that contained the frontmost of the selected objects.

Hot tip

The keyboard shortcut for ungrouping a group is **Ctrl/Command + Shift + G**.

To lock an object or group so that it cannot be accidentally moved, transformed or deleted, select the object or group using the **Selection tool**, and then choose **Object > Lock**. A small Padlock icon appears on the left edge of the object or group to indicate its lock status.

You cannot select a locked object or group.

To unlock a locked object or group, click the **Padlock** icon:

Group separate objects together so that they work as one single unit. Groups are useful when you want to fix the position of particular objects relative to one another. You can move and transform groups without changing the relative position of the individual objects within the group. You can also group two or more groups to form a nested hierarchy.

1 To group objects, make sure that you have two or more objects selected. (See pages 30-31 for techniques for selecting multiple objects.)

2 Choose **Object > Group** (**Ctrl/Command + G**). The objects become a group, and a dotted selection bounding box with eight handles appears, defining the perimeter of the group.

3 To ungroup objects, select the group using the **Selection tool**, then choose **Object > Ungroup**. (If the **Ungroup** command is dimmed, you have not selected a group.) The objects are ungrouped, but all of them remain individually selected. If you want to make changes to an individual object, click on some empty space to deselect the objects, and then reselect the object you want to work on.

4 Use the **Selection tool** to move a group as you would for an individual object. You can constrain the movement vertically or horizontally, you can position a group using X and Y coordinates, and you can also use the **Alt/Option** key to "drag copy" a group. (See page 32 for further instructions.)

Working with Groups

Once you have grouped objects, you can move, scale and transform the group as a single unit. You can also work on individual elements within groups.

Manually Resizing Groups

1 To manually resize a group, select one of the grouped objects using the **Selection tool**. Press and drag on a handle to change the size of all objects in the group. This does not change type size for any text objects in the group.

2 Hold down **Shift**, then press and drag a selection handle to scale objects in the group in proportion. This does not change type size for text objects in the group.

3 Hold down **Ctrl/Command + Shift**, then press and drag a selection handle to scale objects and their contents (type or image) in proportion.

You can also use the Select Previous/Next Objects buttons in the Control panel to cycle through individual objects in a group. Start by selecting an object in a group using the **Direct Selection tool**, then start clicking either the **Select previous object** or **Select next object** button:

4 To select and manipulate an object within a group without having to first ungroup the group, select the **Selection tool** then double-click the object. You can now move the object, scale it, rotate it, delete it, and so on.

5 To reselect a group, double-click very carefully on the edge of the group or object.

Aligning Objects

Hot tip

You can align an object to a specific part of a page by selecting an option from the **Align To** pop-up menu before you click one of the alignment buttons:

Alignment is one of the underlying principles of good design. The Align panel provides controls for aligning objects relative to each other vertically and/or horizontally, as well as to specific parts of the page. The top row of icons in the Align panel controls vertical and horizontal alignment.

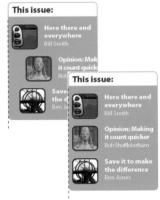

1 To align objects relative to one another, use the **Selection tool** to select two or more objects. Make sure the Align panel is showing by choosing **Window > Object & Layout > Align** (**Shift + F7**) if it is not. Alignment controls are also available in the Align pane of the Properties panel. Make sure **Align to Selection** is selected.

Controlling Alignment

To select a specific object to align on, first select two or more objects:

Then, click once on one of the selected objects. This object is now the key object – a solid highlight appears on the object to indicate this:

When you select an alignment option, the other shapes align relative to the key object:

2 Click one of the **Horizontal Align** buttons to align objects along their left or right edges or horizontal centers. Objects align to the leftmost or rightmost object if you choose **Align left edges** or **Align right edges**. Objects align along the horizontal center point of the selected objects if you choose **Align horizontal centers**.

3 Click one of the **Vertical Align** buttons to align objects along their top or bottom edges or vertical centers. Objects align to the topmost or bottommost object if you choose **Align top edges** or **Align bottom edges**. Objects align along the vertical center point of the selected objects if you choose **Align vertical centers**.

4 Use the Align To pop-up menu to align objects to specific parts of the page/spread.

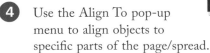

Distributing Objects

To create equal space between objects or equal distance between the lefts, rights, tops or bottoms of selected objects, you can use the Distribute buttons in the Align panel.

The **Distribute options** work only when you have three or more objects selected.

① To distribute objects, select three or more objects using the **Selection tool**.

② Click one of the horizontal **Distribute** buttons to space objects so that the distance from left edge to left edge, right edge to right edge, or horizontal center to horizontal center is equal.

Select the **Use Spacing option** and enter a spacing value before you click on one of the **Distribute Objects** buttons to create a specific amount of space between the edges or centers you specify:

③ Click one of the vertical **Distribute** buttons to space objects so that the distance from top edge to top edge, bottom edge to bottom edge, or vertical center to vertical center is equal.

Distributing Space Equally Between Objects

Rather than spacing objects with equal amounts of space between specific parts of the selected objects, you can create equal amounts of space between each object.

① Choose **Show Options** from the Align panel menu. Two additional buttons appear at the bottom of the panel. Select three or more objects. Click the **Distribute vertical space** button or **Distribute horizontal space** button to create equal amounts of space vertically or horizontally between the selected objects.

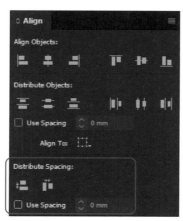

Select the **Use Spacing option** and enter a spacing value before you click one of the **Distribute Spacing** buttons to create a specific amount of space between the selected objects:

Anchored Objects

Anchored objects can be text or graphic frames that are attached to a specific point in text. When the text reflows, the anchored object moves, maintaining its position relative to the point in the text to which it is anchored. The exact position of the anchored object is determined by settings you create in the **Anchored Object Options** dialog box.

Inline Anchored Objects
Inline anchored objects are useful when you want to include a small graphic in the middle of text (like this: ■), and also for larger objects at the start of a paragraph, as in the following example.

To view the **Anchor point markers (¥)** in text, choose **Type > Show Hidden Characters** if they are not already showing.

In the Anchored Object Options dialog box, select the **Prevent Manual Positioning** checkbox to disallow any manual adjustments to the object using the Selection tool.

Above Line anchored objects can be useful when you want an object to behave like a paragraph element and to flow with text.

Select the **Above Line** radio button. Use the Alignment pop-up menu to choose a standard alignment for the object. Enter values for Space Before/Space After to control the amount of space above and below the object:

1 To set up an Inline anchored object, start by creating an object you want to anchor and scaling it to the size at which you want to use it. Choose **Edit > Cut** to place the object on the Clipboard.

2 Select the **Type tool**, then click in the text to place the Text Insertion Point where you want to insert the anchored object.

3 Choose **Edit > Paste** to insert the anchored object at the insertion point. Depending on the size of the object, it may obscure surrounding text.

4 To control the positioning of the object, select it using the **Selection tool**, and then choose **Object > Anchored Object > Options**. With the Inline radio button selected, use the **Y Offset** entry box to control the vertical positioning of the anchored object. Use a negative value to move the object downward.

5 Apply Text Wrap to the object, if necessary, to control the space between the object and the surrounding text.

Irit nibh eu faccum etuerillaore eugait volorero ex ero consed te feu feu feu facin utat. Lent nulput at. Ut velit, con henis sl utpat.

ller in us all
The traveller has existed, in us all, from the dawn of time. Without travel the human being would not be human. Like whales that roam oceans, we must travel, must roam to fulfil a wanderlust that will eventually lead us home.

Anchored Object Options

Position: Inline or Above Line

● Inline
Y Offset: -10pt

○ Above Line
Alignment: Left
Space Before: 0 mm
Space After: -3.528 m

☐ Prevent Manual Positioning

The traveller in us all
The traveller has existed in us all, from the dawn of without travel the human being would not be human. Like whales that roam oceans, we must travel, must roam to fulfil a

The traveller in us all
The traveller has existed, in us all, from the dawn of time. Without travel the human being would not be human. Like whales that roam

8 Working with Color

This chapter shows you how to create, apply and manage color in your documents, using the Color, Swatches and Gradient panels.

Filling and Stroking Objects

Using the **Fill** and **Stroke** boxes, in conjunction with the Swatches panel, you can apply a fill and/or stroke color to a basic shape (such as a rectangle or circle), a text or graphic frame, or a path created with the **Pen** or **Pencil tool**.

When you create a shape – such as a closed path, a rectangle, a circle or a polygon – it is automatically filled with the currently selected fill color, and its path is stroked or outlined with the currently set stroke color and weight.

You can use the **Eyedropper tool** to copy fill and/or stroke attributes from one object to another. Select a "target" object with attributes you want to change. Select the **Eyedropper tool**, then click on the "source" object that has the fill and stroke attributes you want to copy. Its attributes are immediately applied to the target object. Select any other tool in the Toolbox to end the procedure.

When you click on either the **Fill** or the **Stroke box**, its icon comes to the front, indicating that it is now the "active" icon. If you then click on a color in the Swatches panel, you apply the color to the selected attribute – fill or stroke – for the selected object.

If you are working in the Essentials workspace, to apply a fill color to a selected object, click the **Fill box** in the **Appearance pane** of the Properties panel, to access color swatches and related color controls such as **Apply to Frame/Text**. To apply a stroke color, click the **Stroke** box to access color swatches.

Click the color swatch you want to apply.

1. To apply a fill color to a selected object or frame, click on the **Fill** box to make it active. Click on a color swatch in the Swatches panel. (Choose **Window > Swatches** if the Swatches panel is not showing.) The color is applied to the selected object.

2. To apply a stroke color to a selected object or frame, click on the **Stroke** box to make it active. Click on a color swatch in the Swatches panel. The color is applied as a stroke to the path of the selected object.

3. Changing the fill and/or stroke color for a selected object does not change the default fill/stroke color. The default fill color for basic shapes is None, with a 1-point black stroke. To set a default fill/stroke color for objects, make sure nothing is selected, click the **Fill** or **Stroke** box to select it, and then click on a color in the Swatches panel. The **Fill/Stroke** box changes to reflect the color swatch you clicked on. Any basic shapes, or paths drawn with the **Pen** or **Pencil tool**, are automatically filled/stroked with the new default fill/stroke color.

4 To apply a fill or stroke of None to a selected object, path or frame, make sure either the **Fill** or **Stroke** box is selected as required. Click the **Apply** button to reveal the **Apply** pop-up menu if you are working with a single-row Tool panel then click **Apply None**, or simply click the **None** button if you are working with a double-column Tool panel. A red line through the **Fill** or **Stroke** box indicates a fill or stroke of None. Objects with a fill of None are transparent. You can also click on the **None** button in the Swatches panel or in the bottom-left corner of the Color panel.

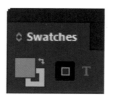

5 Click the **Swap** arrow to swap the fill and stroke colors for a selected object, or hold down **Shift** and press **X** on the keyboard.

6 To apply a default fill of None and a black stroke to a selected object, click the **Default Fill and Stroke** button below the **Fill** box in the Tool panel, or press **D** on the keyboard.

7 The Swatches panel has a miniature representation of the **Fill** and **Stroke** box at the top. This provides a very convenient alternative for making either **Fill** or **Stroke** active, as it is located in the same panel as the color swatches with which you are working.

Coloring Text

1 To apply color to text, select the text with the **Type tool**, make sure that the **Fill** box is selected, and then click on a color swatch in the Swatches panel.

2 When you are working with a text frame selected with the **Selection tool**, you can select the **Formatting Affects Container** button if you want to color the frame's background or stroke. Select the **Formatting Affects Text** button if you want to color the type inside the frame. You also need to make sure that you select the **Fill** or **Stroke** box, as required.

Don't forget The spellings in InDesign are localized. "Color" appears as "Colour" in the UK.

Beware Keyboard shortcuts such as "D" and "X" work only if you do not have the Text Insertion Point active in a text frame.

Beware Be careful when you have a text frame selected with the **Selection tool** and you are applying color. Make sure you have the correct **Formatting Affects Container/Text button** selected, depending on what you want to color.

Hot tip The **Formatting Affects Container/Text buttons** appear in the Tool panel below the **Fill** and **Stroke** boxes, at the top of the Swatches panel and on the left of the Color panel.

The Swatches Panel

 The keyboard shortcut to show/hide the **Swatches** panel is **F5**.

 If you are working in the Essentials workspace, you can access the Swatches panel by clicking either the **Fill** or **Stroke** box in the Appearance pane of the Properties panel:

 A **process color** is printed using the four process inks – Cyan, Magenta, Yellow, and BlacK (CMYK).

 A **spot color** is printed with a premixed ink on a printing press. At 100% (i.e. no tint), a spot color is printed as a solid color and has no dot pattern. Spot colors create their own separate plate when you print separations.

Use the Swatches panel to create a palette of colors you want to use consistently throughout a document. The colors you create and store in the Swatches panel are saved with the document.

The Swatches panel consists initially of a set of default swatches. You cannot make changes to **[None]**, **[Black]** or **[Registration]**. You cannot delete **[None]**, **[Paper]** or **[Registration]**.

Viewing Swatches

1. To show the Swatches panel, choose **Window > Swatches**, or click the **Swatches** icon if the panel is docked in the Panel Dock. Click the **Swatch Views** pop-up at the bottom of the Swatches panel and select an option to control which types of swatches are visible in the panel. You can choose **All Swatches**, **Color Swatches**, **Gradient Swatches** or **Color Groups**.

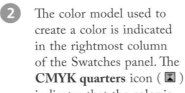

2. The color model used to create a color is indicated in the rightmost column of the Swatches panel. The **CMYK quarters** icon (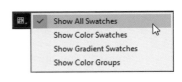) indicates that the color is in CMYK mode. Three bars (■) – Red, Green, Blue – indicate that the color is an RGB color.

3. A gray box to the left of the color model box indicates a process color – a color that will be separated into its Cyan, Magenta, Yellow and Black components when the page is color-separated at output. A gray circle to the left of the color model box indicates a spot color – a color that will create its own plate when printing separations.

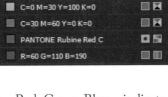

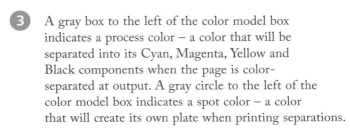

Adding, Deleting and Editing Swatches

Use the Swatches panel to create new process and spot colors, to convert spot colors to process and vice versa, to delete colors, and to create tints and gradients.

Creating New Color Swatches

1 Choose **New Color Swatch** from the Swatches panel menu (▤). Choose either **Spot** or **Process** from the **Color Type** pop-up. Choose one of **CMYK**, **RGB** or **LAB** from the **Color Mode** pop-up, depending on your output requirements.

RGB color **mode** is most useful when you are creating publications that will be printed in-house as color composites, or for screen-based presentations intended for output as PDF, or for the web or multimedia.

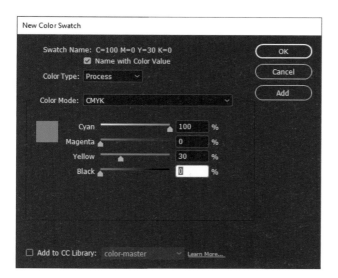

Leave the **Name with Color Value** option selected for InDesign to automatically name the color with the exact color breakdown you define. Switch off the option and enter a name if you want to name the color yourself.

2 Drag the color slider markers (◨) or enter values in the % entry fields. If you have an object selected when you create a new color swatch, the new color is applied to either its fill or its stroke, depending on whether the **Fill** or **Stroke** box is active. If no object is selected when you create a new color, it becomes the new default color for fill or stroke, again depending on which box is active.

Deleting Swatches from the Swatches Panel

1 To remove a swatch from the Swatches panel, click on it to select it, and then click the **Trash** button. Alternatively, drag the swatch onto the Trash. In the **Delete Swatch** dialog box, use the **Defined Swatch** pop-up to choose a color (from the remaining colors) that will replace instances of the color you are deleting.

Select **Unnamed Swatch** to leave objects in the document that are already filled or stroked with the color you are deleting filled or stroked with that color. The color will no longer appear in the Swatches panel as a "named" swatch.

...cont'd

To convert a color **from Spot to Process** or vice versa, double-click the swatch and then use the **Color Type** pop-up to change from one type to the other.

If you do not have anything selected when you edit a swatch, all objects to which the color was previously applied are updated, and the edited color becomes the default fill or stroke color, depending on which color box is active in the Tool panel.

Tints of a spot color print on the same separations plate as the spot color. A tint of a process color multiplies each of the CMYK process inks by the tint percentage. For example, a 50% tint of C=0 M=40 Y=100 K=10 creates a tint color of C=0 M=20 Y=50 K=5.

Editing Existing Colors

1 Click on the color swatch to select it, and then choose **Swatch Options** from the Swatches panel menu (▦); alternatively, double-click the swatch you want to edit.

2 Use the **Swatch Options** dialog box to make changes to the color. Select the **Preview** option to see the changes implemented in the document before you **OK** the dialog box. Click **OK** when you are satisfied. If you have an object selected when you edit a color swatch, the fill or stroke of the object will change to reflect the change made to the color. All other objects to which the color has been previously applied will also update accordingly.

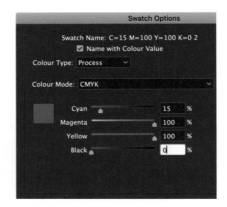

Creating Tints

You can create tints of existing spot or process colors using the Swatches panel menu.

1 To create a tint, click on a color in the Swatches panel to set it as the base color. Choose **New Tint Swatch** from the Swatches panel menu. Drag the Tint slider (◣), or enter a percentage value to define the tint. **OK** the dialog box. The tint appears in the Swatches panel with the same name as the original base color, but with the tint % value also indicated.

2 If you edit a color that is also the base color for a tint, the tint is adjusted accordingly.

The Color Panel

The Swatches panel is the primary panel for creating and editing colors in Adobe InDesign. You can also mix colors in the Color panel and then save the color as a swatch so that it becomes a "named" color.

To show the **Color panel**, choose **Window > Color** or click the **Color** icon if the panel is in the Panel Dock, or use the keyboard shortcut **F6**.

1 If you have a color currently selected in the Swatches panel, the Color panel initially appears with the Tint slider visible and displays the selected color. To create a color in the Color panel, click either the **Fill** or the **Stroke** box, then choose a color model from the panel menu.

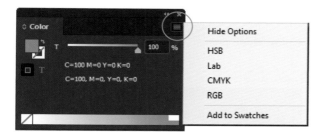

If you select an object and then edit its fill or stroke color in the Color panel, the change is only applied to the selected object; the change does not affect other objects to which the original color is applied.

2 Either drag the color component sliders (▲) or enter values in the entry fields. You can also select **None**, **Black** or **White**, or click in the color spectrum bar at the bottom of the panel.

3 To save a color created in the Color panel, click the **New Swatch** button (▣) in the Swatches panel, or choose **Add to Swatches** from the Color panel menu (▤). The color becomes a swatch in the Swatches panel and is now a named color. The Color panel displays the Tint slider for the color, as it is now a swatch.

4 A "named" color is a color that has an entry in the Swatches panel. An "unnamed" color is one that you have created using the Color panel, and possibly applied to an object in your document, but that does not appear in the Swatches panel. It is easier to identify, edit and manage colors if they appear as named colors in the Swatches panel. Creating named colors from the outset is a good habit to get into.

A "named" color is a color that appears in the Swatches panel. As such, it is saved with the document and can be used repeatedly and consistently. Choose **Add Unnamed Colors** from the Swatches panel menu to create named swatches for all unnamed colors in the document.

The **Web color library** helps guarantee consistent color results on both Windows and Macintosh platforms. The Web panel consists of the 216 RGB colors most commonly used by web browsers to display 8-bit images.

Color Matching Systems

Color matching systems such as PANTONE®, FOCOLTONE® and TRUMATCH™ are necessary when you want to reproduce colors accurately, especially colors used for corporate branding, which need to be reproduced consistently. You can choose predefined colors from a number of color-matching systems.

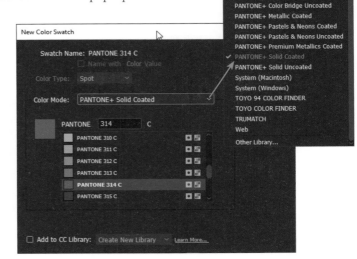

1 To choose a color from a color-matching system such as PANTONE, show the Swatches panel. Choose **New Swatch** from the panel menu (▤).

2 In the **New Color Swatch** dialog box, select a color-matching system from the **Color Mode** pop-up menu.

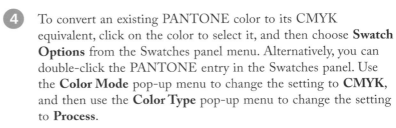

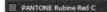

You can identify spot colors in the Swatches panel by the **Spot Color icon** that appears to the right of the spot color entry in the Swatches panel:

3 Either scroll through the PANTONE list or, to access a color swatch quickly, type the number of the PANTONE color you want to select into the PANTONE box. Click **OK** to add the color to the Swatches panel.

4 To convert an existing PANTONE color to its CMYK equivalent, click on the color to select it, and then choose **Swatch Options** from the Swatches panel menu. Alternatively, you can double-click the PANTONE entry in the Swatches panel. Use the **Color Mode** pop-up menu to change the setting to **CMYK**, and then use the **Color Type** pop-up menu to change the setting to **Process**.

Creating and Applying Gradients

A gradient is a gradual transition from one color to another color. Gradients can be linear or radial. You can fill objects and frames with a gradient fill, and they can be applied to strokes and also to text – without first having to convert the text to paths.

To **add additional colors** to a gradient, position your cursor just below the Gradient Ramp, and then click to add another color stop. Apply color to additional stops in the same way that you apply color to the Start and End stops:

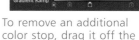

To remove an additional color stop, drag it off the Gradient Ramp.

1 To create a gradient, choose **New Gradient Swatch** from the Swatches panel menu (▤). Enter a name for the gradient.

2 Use the **Type** pop-up to choose between **Linear** and **Radial**. To specify the start and end colors for the gradient, click on either of the "stop" icons on the Gradient Ramp. The triangle on the top of the stop becomes highlighted (▣) to indicate that it is selected.

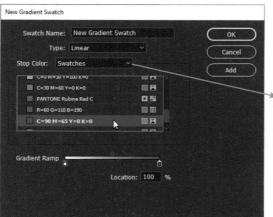

3 Choose **Swatches** from the **Stop Color** pop-up menu to apply an existing color swatch. Alternatively, enter CMYK values or drag the sliders to mix a color. Click on the other **Stop** icon and repeat the process.

When you work with the **Gradient panel**, make sure that the **Swatches panel** is visible so that you can choose start and end colors for the gradient.

4 Drag the diamond slider (◆) along the top of the Gradient Ramp to control the point at which both colors in the gradient are at 50%. Click **OK** when you are satisfied with the settings. The gradient is added to the Swatches panel.

The Gradient Panel

You can also use the Gradient panel to create a gradient.

Click the **Reverse button** if you want the gradient to flow in the opposite direction:

1 Choose **Window** > **Color** > **Gradient** to show the Gradient panel,

...cont'd

Radial gradients start from the center and work outward.

or click the **Gradient** icon if the panel is docked in the Panel Dock.

2 Choose from **Linear** or **Radial** in the **Type** pop-up.

3 To specify the start and end colors for the gradient, click a **Stop** icon on the Gradient Ramp to select it. Hold down **Alt/Option**, and click on a color in the Swatches panel. Repeat this process for the other stop.

4 Enter an angle for the gradient in the **Angle** entry field.

5 Click the **New Swatch** button (⊞) in the Swatches panel to add the gradient swatch to the Swatches panel.

Hold down **Shift** as you drag with the **Gradient tool** to constrain the angle of the gradient to multiples of 45 degrees.

Applying Gradients

Once you have saved a gradient in the Swatches panel, you can apply it to objects, frames, strokes and text.

1 To apply a gradient fill, select an object or frame. Click the **Fill** box in the Toolbox to make it active. Then, click on a gradient swatch in the Swatches panel. Alternatively, select **Apply Gradient** from the **Apply** pop-up menu below the **Fill/Stroke** boxes in the Tool panel to apply the most recently selected gradient.

2 The **Gradient tool** (▇) allows you to control the angle and length of a gradient. Make sure you select an object with a gradient applied to it. Select the **Gradient tool**.

3 Position your cursor on the gradient object, then press and drag. As you do so, you will see a line. The line determines the direction and length of the gradient. For a linear gradient, the start and end colors fill any part of the object you do not drag the line across. For radial gradients, the end color fills any part of the object you do not drag the line across.

To **apply a gradient to text**, either select the text with the **Type tool** or select a text frame with the **Selection tool**, but make sure that you also select the **Formatting Affects Text button** (▢T) in the Tool panel or the Swatches panel. Click on a gradient swatch, or click the **Apply Gradient** button:

9 Managing and Editing Text

This chapter covers the Check Spelling dialog box, the Find and Replace dialog box, type on paths, as well as the Text Wrap panel.

Spell Checking

To control the range of errors that the **Spelling** dialog box identifies, choose **Edit** > **Preferences** > **Spelling** (Windows), or **InDesign** > **Preferences** > **Spelling** (Mac). Deselect options you want the spell check to ignore.

Click the **Add button** to indicate that a word is spelled correctly and add it to the user dictionary.

Before you begin spell checking, make sure the correct language is selected – choose **Edit** > **Preferences** > **Dictionary** (Windows) or **InDesign** > **Preferences** > **Dictionary** (Mac), then choose the correct language from the Language pop-up.

1. To check spelling in a document, select the **Type tool**, and then either highlight a range of text or click into a text frame to place the Text Insertion Point. Choose **Edit** > **Spelling** > **Check Spelling**.

2. Choose an option from the **Search** pop-up to define the scope of the spell check, and click the **Start** button. InDesign highlights the first word not in its spell check dictionary. The unrecognized word appears in the **Not in Dictionary** field and in the **Change To** field. InDesign lists possible correct spellings in the **Suggested Corrections** list box.

3. To replace the incorrect word with a suggestion from the list, click on a suggested word, and then click the **Change** button. InDesign substitutes the correction in the text and moves on to the next unrecognized word. Click the **Change All** button to change every instance of the same spelling error. To accept the spelling of an unrecognized word as correct, click the **Ignore** button. Click the **Ignore All** button if there are multiple instances of the word in the story.

4. Click the **Done** button to finish spell checking, either when InDesign has checked the entire story or at any time during spell checking.

Choose **Edit** > **Spelling** > **Dynamic Spelling** if you want InDesign to highlight possible misspelled words with a red underline. Use the Context menu to make corrections.

Adding Words to the Dictionary

You can add words that InDesign does not recognize to the selected dictionary. This is particularly useful if you regularly use specialist or technical words in your documents.

1 To add a word to the user dictionary, first start a spell check. When InDesign identifies a word not in the dictionary that you want to add, click the **Add** button.

The Dictionary Command

You can use the **Dictionary** dialog box to add and remove words at any time as you work on your document.

1 To add or remove a word, highlight a word then choose **Edit > Spelling > User Dictionary**.

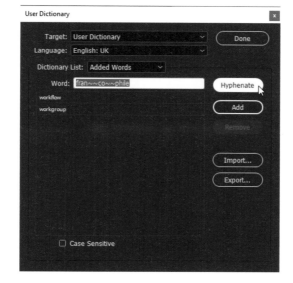

2 Use the default **User Dictionary** as the target dictionary to store hyphenation and spelling exceptions in a dictionary file that resides on the hard disk of your computer. Select a document name from the **Target** pop-up menu if you want to store spelling and hyphenation exceptions inside the document.

3 Click the **Hyphenate** button to see InDesign's suggested hyphenation breaks, indicated by tilde marks (~).

4 Click the **Add** button to add the word to the dictionary. The word appears in the list box with its hyphenation points indicated. To remove a word from the list, click on a word in the list and then click the **Remove** button. Click **Done** when you have finished adding and removing words.

Use **Dictionary Preferences** to specify whether the text-composition engine composes text using the word list from the user dictionary, the document's internal dictionary, or both.

You can override InDesign's hyphenation suggestions by inserting your own tilde marks. Enter one, two, or three tilde marks (~) to rank hyphenation points. One tilde mark indicates your preferred choice. Three tilde marks represents your least preferred choice. Enter the same number of tilde marks to indicate equal ranking.

To prevent all instances of a word from hyphenating, enter a tilde mark in front of the word.

Finding and Changing Words

The **Find/Change** dialog box allows you to find particular words or phrases in a story or document and then change them to something else – for example, you could change a misspelling of a technical word throughout a story or an entire publication.

Use the pop-up to the right of the **Find what** and **Change to** entry fields () to search for, or replace with, special characters.

1 To find and change one word or phrase to another word or phrase, select the **Type tool**. Click into a text frame to place the Text Insertion Point. Choose **Edit > Find/Change**. Enter the word or phrase you want to search for in the **Find what** entry field. Enter the text you want to change to in the **Change to** entry field. Use the **Search** pop-up to specify the scope of the Find/Replace routine.

Use the **Search option buttons** to extend searching to locked/hidden layers, master pages and footnotes:

2 Select **Whole Word** (▤) to ensure that InDesign finds only instances of complete words. For example, if you search for "as", select the **Whole Word** option so that the search does not find instances of the a + s character pair in words such as "was" and "class".

Select the **Case Sensitive** option (**Aa**) when you want to find text that matches exactly the capitalization used in the **Find what** field. For example, if you entered "InDesign" with the **Case Sensitive** option selected, the search would not find "indesign" or "Indesign". Also, when you use the **Case Sensitive** option, the text you change to matches the exact capitalization of the text in the **Change to** field.

3 When you have made the appropriate selections and entered the **Find what** and **Change to** text, click **Find Next**. InDesign moves to the first instance of the Find text and highlights it, scrolling the document window if necessary to show it. Click the **Change** button to change that instance only. Click the **Find Next** button to continue the search.

4 Click the **Change/Find** button to change the highlighted text and move to the next instance. Click **Change All** to change every instance of the **Find what** text. A **Search Complete** box indicates how many instances were changed.

Finding and Changing Basic Text Formatting

You can also use the **Find/Change** dialog box to search for instances of formatting attributes, such as a particular font, size or style, and then change these attributes to something different. You can also find and replace paragraph and character styles.

1 To search for text formatting attributes, select the **Type tool**, and click into a text frame to place the Text Insertion Point, or highlight a range of text if you want to limit the operation to specific text.

2 Choose **Edit > Find/Change**. Use the **Search** options to define the extent of the Find/Change routine. Do not enter any text in the **Find what/Change to** entry fields. Click the **More Options** button (which becomes **Fewer Options**).

3 Click the **Specify attributes to find** button ( ) in the **Find Format Settings** area. Select a category from the Categories list box on the left, and then select the attributes you want to find. Make sure you specify exactly the attributes you want to search for. If you leave a field blank, this indicates to InDesign that the attribute is not relevant to your search. For example, if you want

to find instances of Arial Bold in a story, you can leave the **Size** entry field blank; InDesign will then find any instance of Arial Bold, regardless of its size. Enter a size value only if you want to limit the search to a particular character size.

Use the **Grep tab** to perform complex, pattern-based searches on text and formatting. The **Glyph tab** allows you to find and replace glyphs. Use the **Object tab** to search for and replace attributes and effects applied to objects.

Be logical and patient when finding and changing complex sets of attributes. It is easy to accidentally specify an attribute that does not exist in your publication, and consequently get a message that no instances were found.

The more values you enter and/or options you choose, the more limited the search becomes.

...cont'd

There are Find/ Change capabilities for virtually all character and paragraph formatting attributes. Click the various categories in the **Change Format Settings** dialog box to explore the possibilities, or select existing paragraph or character styles from the pop-up menus in the **Change Format Settings** dialog box.

④ **OK** the dialog box when you are satisfied with the settings. The settings you have chosen are indicated in the **Find Format Settings** list box.

⑤ Click the **Specify attributes to change** button ( ) in the **Change Format** area. Select a category from the Categories list box, and select the formatting attributes you want to change to. Make sure you specify exactly the attributes you want. **OK** the dialog box when you are satisfied. The settings you chose are indicated in the **Change Format** readout box. When you specify Formats, warning icons appear next to the **Find what/ Change to** fields to indicate that

formatting settings are in force. These are especially useful if you have clicked the **Fewer Options** button to condense the panel, as unnecessary Format settings can cause simple Find/Change operations to go wrong.

Typically, when things go wrong with a Find/Change operation it is because the attributes you specify for the search do not exist in the document. Go back out of the Find/ Change dialog box and check that you know exactly what you are looking for by checking actual settings on text in the story.

⑥ When you no longer need your Format settings, click the **Clear attributes** buttons (🗑) in the **Format Settings** area to revert all **Find/Change** settings to their default values. If you are having problems getting a particular Find/Change routine to work, it is sometimes useful to use the **Clear** button to reset everything, and then set up your Find/Change criteria from scratch.

⑦ Use the **Find Next**, **Change**, **Change All** and **Change/Find** buttons to proceed with the search-and-replace operation.

Text Wrap

Controlling Text Wrap becomes necessary when you start to combine text frames and graphic frames on a page, particularly when they overlap each other.

If you no longer need Text Wrap settings on an object, make sure it is selected, and then click the **No Text Wrap button** in the Text Wrap panel:

1 To wrap text around a graphic frame, select the graphic frame using the **Selection tool**. It doesn't matter whether the picture frame is in front of or behind the text frame.

2 Choose **Window** > **Text Wrap** (**Ctrl/Command + Alt/Option + W**) to show the Text Wrap panel. Click the **Wrap around bounding box** button to wrap text around all sides of the graphic frame.

3 Enter Offset values for Top, Bottom, Left and Right to control how far text is pushed away from the various edges of the frame. Press **Enter/Return** to apply changes. When

you select the frame with the **Selection tool**, a faint blue standoff border appears around the frame to visually indicate the Text Wrap area (provided the frame edges are visible).

4 Click the **Jump Object** button to prevent text flowing on either side of the picture frame. When you choose this option you can control only the Top and Bottom offset amounts.

If you are working in the **Essentials workspace**, you can access Text Wrap controls in the Text Wrap pane of the Properties panel.

You can click the **More options button** to reveal/hide the full set of Text Wrap controls.

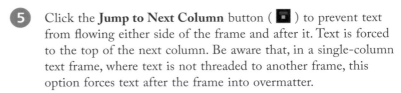

...cont'd

When wrapping around a bounding box or object shape, use the **Wrap To pop-up menu** to control whether the Text Wrap effect applies to specific sides of the object.

5 Click the **Jump to Next Column** button ( ) to prevent text from flowing either side of the frame and after it. Text is forced to the top of the next column. Be aware that, in a single-column text frame, where text is not threaded to another frame, this option forces text after the frame into overmatter.

6 You can also apply Text Wrap to a selected text frame. This can be useful for such things as pull quotes, where text frames overlap other text frames.

Text Wrap and Stacking Order

When you place a text frame over an image that has Text Wrap applied to it, some or all of the text may disappear from the frame. This is caused by the **Text Wrap** setting, which affects text in frames both behind and in front of it. Use the following technique to prevent this happening on an individual text frame.

If you prefer **Text Wrap** settings to affect only text in frames below the graphic frame in the stacking order, choose **Edit > Preferences > Composition** (Windows) or **InDesign > Preferences > Composition** (Mac), then select the **Text Wrap Only Affects Text Beneath** option:

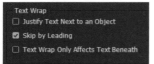

1 Select the text frame using the **Selection tool**. **Choose Object > Text Frame Options (Ctrl/ Command + B)**. Select the **Ignore Text Wrap** option. This affects the individual text frame to which you apply the control.

Irregular Text Wrap

You can also wrap text around non-rectangular shape objects such as circles and stars, and along paths created with the **Pen** or **Pencil tool**. In addition, you can create Text Wrap based on a clipping path, or the shape of an imported Adobe Illustrator graphic.

Wrapping Text Around Shapes

1 To wrap text around a shape frame or a path, choose **Window > Text Wrap (Ctrl/Command + Alt/Option + W)** to show the Text Wrap panel if it is not already showing. Select the object, then click the **Wrap around object shape** button. Enter an Offset value. For non-rectangular objects you create a single, standard offset amount – there is only one **Offset** field available.

See pages 84-86 for information on creating, manipulating and importing images with clipping paths.

Wrapping Text to Clipping Paths

1 To wrap text around a graphic with a clipping path, make sure you select the image with the **Selection tool**. Click on the **Wrap around object shape** button.

2 Choose **Show Options** from the panel menu or click the **Expand** button (◆) in the panel tab to show the extended panel. Choose **Same as Clipping** from the **Type** pop-up menu.

To wrap text around an Adobe Illustrator graphic, select **Detect Edges** from the Contour Options pop-up menu:

3 Enter a value in the **Offset** field to control the distance that the text is offset from the image's clipping path.

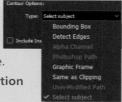

For images that have a well-defined subject, you can use the **Select subject** option in the **Contour Options > Type** menu to create an automated Text Wrap around the identified subject in the image.

If necessary, you can use the **Direct Selection** tool to fine-tune the wrap path.

Type on a Path

Running type along a path can produce interesting and unusual results. You can apply type to open and closed paths, including shapes or frames.

You cannot create type on compound paths, such as those created when you use the Pathfinder commands.

The **Type on a Path tool** is grouped with the **Type tool** (see page 16).

If you enter too much text, the additional text becomes **overmatter**. To see all the text, either thread the text onto another path or into another frame or reduce the type size.

1 To apply type to a path, select the **Type on a Path** tool. Position your cursor on a path. The path does not need to be selected, but make sure you see the additional "+" symbol appear on the cursor ($\mathcal{F}^+$), which indicates that the text will be applied to the path.

2 Click on the path. A Text Insertion Point appears on the path. Enter type using the keyboard. You can enter type along the entire length of the path.

3 When you select the text with a **Selection tool**, the In- and Out-ports appear at the start and end of the path. On the inside of the ports are the start and end brackets. Drag the start or end bracket to adjust the length of the text area on the path.

4 In the middle of the path is the path type center bracket. Drag the center bracket to reposition text along the path after you have adjusted one of the end brackets.

5 To create additional settings for type on a path, with the path type selected choose

Type > Type on a Path > Options. Use the **Align** and **To Path** pop-ups to specify which part of the type aligns to which part of the path. In this example, the ascenders of the type align to the bottom of the path. Use the **Spacing** control to compensate, if necessary, for the way in which characters fan out on some curves.

You can flip type manually by dragging the center bracket across the path, or by selecting the **Flip** checkbox in the **Type on a Path Options** dialog box.

10 The Pages Panel and Master Pages

The Pages panel provides key functionality for adding, deleting and positioning document pages, as well as essential controls for working with master pages.

The Pages Panel

Use the Pages panel to move from page to page, to move to master pages, and to add and delete both document pages and master pages. Click the **Pages** icon if the panel is docked in the Panel Dock, or choose **Window > Pages** (**F12**) to show the panel.

When working in the Pages panel, an initial distinction needs to be made concerning the way in which you select a spread or page on which to work. In InDesign you can "select" or "target" a spread or page.

If you are working in the Essentials workspace, the Pages panel is one of the three default panels in the Panel Dock. Click the **Pages tab** to access the panel.

Selecting Spreads/Pages

Select a page or spread when you want to change settings such as margin and column settings and guides on a particular spread – in other words, settings that affect the page rather than objects on the page. This is most important when there are multiple pages displayed at a low level of magnification in the document window. A selected page is indicated by a highlighted page icon, not highlighted page numbers.

1 To select a page, click once on the page icon in the Pages panel. To select a spread, click once on the page numbers below the spread icons.

2 Double-click a page icon to select and target it. The page or spread is centered in the document window.

A **spread** normally consists of two pages viewed side by side as, for example, in a magazine or this book.

Targeting Spreads/Pages

Target a page or spread when you want to make changes to objects on a particular spread or page. For example, when more than one spread is visible in the document window and you want to paste an object onto a particular spread, make sure you target it before you paste the object.

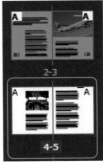

A "targeted" page or spread is indicated by a highlighted page number ( ), as opposed to a highlighted page icon.

1 Working on, selecting, or modifying an object on a page automatically activates the page or spread as the target page or spread.

2 Click on a page or its Pasteboard area in the document window to target the page or spread.

3 Alternatively, in the Pages panel, double-click the page number below the page or spread icon. The pages are centered in the document window.

Pages Panel Options

You can change the arrangement of pages in the Pages panel and the appearance of page thumbnails using options in the **Panel Options** dialog box.

1 To show the Panel Options dialog box, choose **Panel Options** from the Pages panel menu (▤).

2 Use the **Size** pop-up menu for Pages and Masters to control the size of page thumbnails. The larger the thumbnail, the more page layout detail you can pick out. Larger thumbnails can make it easier to identify and navigate accurately to specific pages in a document.

3 To create a horizontal arrangement of page thumbnails in the Pages panel, select **View Pages** > **Horizontally** from the Pages panel menu.

Hot tip

To apply a color label to thumbnails so that you can categorize and identify them easily, select a page or a range of pages, then choose **Page Attributes** > **Color Label** from the Page panel menu. Select a color from the pop-up color menu:

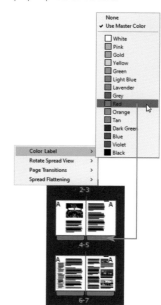

...cont'd

④ Use controls in the **Icons** area of the dialog box to specify whether or not page thumbnails display additional icons to indicate the presence of transparency, spread rotation and page transitions.

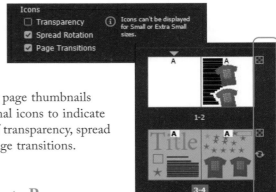

Moving from Page to Page

There are a number of ways you can move from page to page in a multi-page document. You can use the Pages panel, the scroll bars, keyboard shortcuts, or the page indicator area in the bottom-left corner of the document window.

① In the Pages panel, double-click the page icon of the page you want to move to. The page icon you double-click on highlights, as does the page number below it. The page you double-click is centered in the document window.

② To use the page indicator area, either click the **First Page**, **Previous Page**, **Next Page** or **Last Page** button or highlight the **Page Number** entry field, enter the number of the page to which you want to move, and then press **Enter/Return**. Alternatively, use the **Pages** pop-up menu and select a page number to move to.

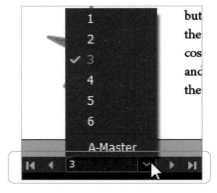

③ To move to other pages, you can use the **Up** and **Down** scroll arrows of the publication window to move to different pages or spreads.

④ You can also use the **Shift + Page Up/Page Down** keyboard shortcut to move backward/forward one page at a time.

Use the keyboard shortcut **Ctrl/Command + J** to access the **Go to Page dialog box**; enter a page number, and then click **OK** to move to that page:

Inserting and Deleting Pages

You can specify the number of pages in a document in the **New Document** dialog box, but you can also add and delete pages in the document at any time.

1 To add a page after the currently targeted page or spread – indicated by the highlighted page numbers in the Pages panel – click the **New Page** button (▣) in the Pages panel. The new page is automatically based on the same master as the currently targeted page. A master page prefix (which is typically a letter) in the document page icon indicates the master page on which it is based.

2 To insert a single page or multiple pages using the **Insert Pages** dialog box, choose **Insert Pages...** from the Pages panel menu (▤). Enter the number of pages you want to add. Use the **Insert Pages** pop-up to specify the placement of the additional pages relative to the page number you specify in the **Page Number** entry field. In a publication with multiple master pages, choose the master on which the additional pages will be based from the **Master** pop-up. **OK** the dialog box.

3 To delete a page, click on the page to select it, and then click the **Trash** button (or you can drag the page onto the **Trash** button). You can also choose **Delete Page/ Spread** from the Pages panel menu. **OK** the warning dialog box.

To select a consecutive range of pages, select the first page in the range then hold down **Shift**, and click on the last page to highlight all pages between the two clicks:

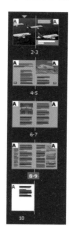

To select non-consecutive pages, select a page then hold down **Ctrl/Command**, and click on other pages to add them to the selection:

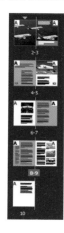

Setting Master Pages

In multi-page documents, position those elements such as automatic page numbering, datelines, headers and footers, and logos that you want to appear on all – or nearly all – of the pages in the document on the master page. When you add pages based on a master page to a document, all the objects on the master page are automatically displayed on the document pages. Master pages are essential for guaranteeing that such objects repeat consistently throughout the document. This is also the most efficient method for placing such repeating objects.

If you edit objects on a master page, these changes are automatically applied to master page objects appearing on all document pages based on that master, provided that you have not edited individual instances of the objects on the document pages.

Each new document you create has an **A-Master** page by default. All document pages are initially based on the A-Master. When you start a new document, the **A-Master** settings are determined by the margin and column settings that you specify in the **New Document** dialog box.

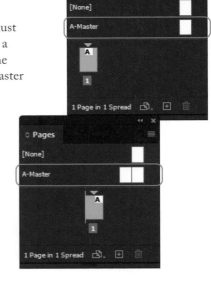

1 To set a master page, you must first move to it. To move to a master page, double-click the **Master Page** icon in the Master Pages section of the Pages panel. In a single-sided publication, **[None]** and a single **A-Master** appear by default as soon as you create the publication. In a facing-pages publication, **[None]** and a double-sided **A-Master** appear by default.

If you select the **Primary Text Frame option** in the **New Document** dialog box, a text frame is automatically added to the A-Master. The text frame fits to the specified margins and matches the number of columns you specify:

A primary text frame appears on all pages based on the A-Master.

2 Alternatively, choose **Layout > Go to Page (Ctrl/Command + J)**, then in the **Page** entry box, type in the prefix ("**A**" for an A-Master, "**B**" for a B-Master, and so on). Click **OK** to move to the specified master page.

3 Create, position and manipulate text, graphic and shape frames, lines and paths on the master page as you would on any document page. Objects you create on master pages have a dotted line to indicate the frame edge. This helps differentiate master page objects from objects you place on document pages.

Hot tip

Objects placed on master pages cannot be selected initially on the document pages where they appear. (See page 132 for information on making these master page elements editable.)

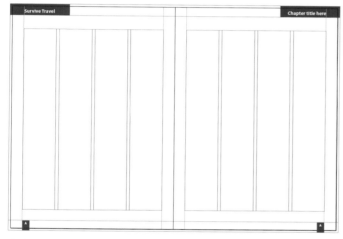

Beware

In a publication with layers, master page objects appear behind other objects that are on the same layer on document pages.

4 When you need to edit objects placed on a master page, return to the master page, and then make changes to objects as necessary. Any changes you make are updated on all document pages based on that master page.

5 Set up ruler guides on a master page if you want them to appear on all publication pages based on that master page. Ruler guides that you create on a master page cannot be edited on a document page unless you create a local override on the guide. (See page 132 for information on creating overrides on master page objects.)

6 In the **New Document** dialog box, select the **Primary Text Frame** checkbox to automatically create a text frame on the A-Master page. The primary text frame

Pages	Facing Pages
12	☑
Start #	Primary Text Frame
1	☑

fits within the margins and has the same number of columns and gutter width as defined in the **New Document** dialog box.

Beware

If you switch on **Layout Adjustments** and then apply a different master to a document page, the position and size of objects may alter on the document page.

Automatic Page Numbering

Automatic page numbering is useful in multi-page documents that need to have sequential page numbering. Set up automatic page numbering on a master page to automatically number all of the document pages based on that master.

To apply **automatic page numbering** only to individual document pages, move to the document page and then use the same procedure as for setting automatic page numbering on a master page.

1 To set up automatic page numbering, double-click the **Master Page** icon in the Pages panel. The icon becomes highlighted, indicating that you are now working on the master page. Also, the page indicator in the bottom-left corner of the publication window indicates that you are now working on a master.

2 Create and position a text frame where you want page numbers to appear on all pages based on the current master. (See Chapter 3 for information on creating text frames and entering text.)

3 Make sure the Text Insertion Point is flashing in the frame. Choose **Type** > **Insert Special Character** > **Markers** > **Current Page Number** (**Ctrl/Command** + **Alt/Option** + **N**). Alternatively, you can click the right mouse button (Windows), or **Ctrl** + **Click** (Mac), to access the context-sensitive menu. Choose **Insert Special Character** > **Markers** > **Current Page Number**. An "A" appears in the text frame. This is the automatic page number symbol for A-Master pages.

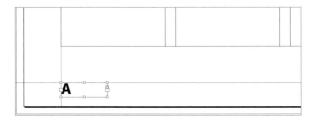

Placing the automatic page numbering symbol on a master page guarantees exactly the same positioning and formatting for page numbers throughout a document.

4 Highlight and format the "A" as you would for any other text character. Add a prefix such as "page" or suffix such as "of 20" as necessary.

5 Move to a document page based on the A-Master to see the automatic page numbers appearing on document pages.

6 If you are setting up a facing-pages document, remember to set up automatic page numbering on both the left- and right-hand master pages.

Sectioning a Document

When you are working with automatic page numbering and you want to change the numbering of a specific range of pages, you can create a section in a document. For example, in a magazine production environment, you might want a feature spread to begin on page 28, not page 2.

1 To change page numbering in a document by creating a section, double-click the page icon for the page where you want to create a section.

2 Choose **Numbering & Section Options** from the Pages panel menu (▤).

3 Select the **Start Page Numbering at** checkbox and enter the page number where you want to begin the section. **OK** the dialog box.

4 The page number indicator below the page icon changes to the number you specify. All subsequent page numbers after the sectioned page are numbered sequentially. The section is identified by a triangle above the section page icon.

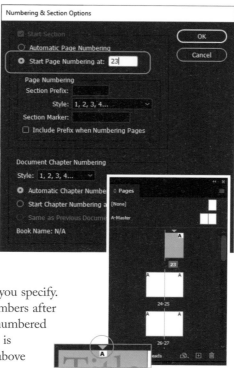

The **Allow Pages to Shuffle option** in the Pages panel menu controls how pages and spreads are rearranged when you add, remove or section pages in a document.

In a facing-pages document, if you section an even-numbered page but enter an odd number, provided that **Allow Pages to Shuffle** is selected in the Pages panel menu, the left-hand page becomes a right-hand page and subsequent pages and their contents shuffle: right-hand pages become left-hand pages (and vice versa).

Overriding Master Page Objects

By default, master page objects cannot be edited or manipulated on document pages. You can override master page objects if you want to make changes to them on individual document pages – for example, you might need to amend a running header set up on a master to indicate a particular section of a publication.

After you create an override on a master page object, you can modify its attributes, such as fill and stroke, position and size, as well as its contents – image or text. When you modify a particular attribute, that attribute is no longer associated with the original attribute on the master page object. The attribute no longer updates if you change the original attribute on the master page object. Attributes you do not modify remain associated with their equivalent attribute on the master page object and update if you change them on the master page object.

1 To override a master page object on a document page, working with the **Selection tool**, hold down **Ctrl/ Command + Shift** and click on the object. It is no longer a protected

master page object; the frame edges of the object change to solid and it can be edited and manipulated as any other object on the page. InDesign refers to this process as creating a "local override".

2 To create overrides on all master page objects on a page/spread, double-click on the page number(s) in the Pages panel to target the page/spread, and then choose **Override All Master Page Items** from the Pages panel menu (▤).

3 To remove an override from a specific object only, select the object then choose **Remove Selected Local Overrides** from the panel menu.

4 If you want to remove all local overrides, double-click on the page number(s) in the Pages panel to target the page/spread then choose **Remove All Local Overrides** from the panel menu. This removes local overrides from all objects on the page or spread.

5 To hide all master objects on a page/spread, target the page/ spread and choose **View > Hide Master Items**. Choose **View > Show Master Items** to make them visible again.

You can **detach a master page object** from its master page so that it is completely disassociated from the original master page object. To detach a master page object, first create an override for the object, and then choose **Detach Selection From Master** from the Pages panel menu.

11 Paragraph and Character Styles

Paragraph and character styles deliver speed, efficiency and – above all – consistency to the appearance of your InDesign documents.

Paragraph Styles

The **Paragraph Styles** and **Character Styles** settings help guarantee consistency within a single document, through a series of related documents (for example, the chapters in a book) and through a series of publications (such as different books in the same series).

A paragraph style is a collection of character and paragraph attributes that can be given a name (e.g. Body1) and saved. Once saved, the paragraph style can be quickly and easily applied to individual paragraphs or ranges of paragraphs, guaranteeing consistency of formatting and also speeding the process of styling text. Paragraph styles control the character and paragraph settings for complete paragraphs.

 To create a paragraph style, choose **Type** > **Paragraph Styles** (**F11**) to show the Paragraph Styles panel, or click the **Paragraph Styles** icon if the panel is in the Panel Dock. Choose **New Paragraph Style** from the Paragraph Styles panel menu (▤).

134

If you are working in the Essentials workspace, you can access paragraph styles from the **Paragraph Styles** pop-up in the **Text Style pane** of the Properties panel:

As a general rule of thumb, paragraph and character styles are worth creating if you intend to use the same settings more than a couple of times in the same document, or if you want to keep settings consistent across more than one publication.

 Enter a name for the paragraph style. With **General** selected in the formatting categories list, leave the **Based On** pop-up set to **[No Paragraph Style]** and the **Next Style** pop-up as **[Same style]**. To enter a shortcut, make sure **Num Lock** is **On**. Hold down any combination of **Ctrl/Command**, **Alt/Option**, and **Shift**, and press any key on the numeric keypad. Setting a shortcut for the style allows you to apply the style without using the mouse when you are styling your text.

...cont'd

3 Click on Formatting category options in the list box on the left of the dialog box to choose the categories for which you want to define settings. Settings available in the various categories are covered in detail in individual chapters dealing with character and paragraph formatting, tabs and rules. The most important choices initially are **Basic Character Formats** (to set attributes such as Font, Size and Leading) and **Indents and Spacing** (to set alignment options, indents, and the space before and after paragraphs).

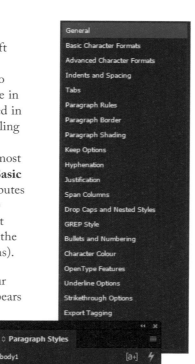

4 When you are satisfied with your settings, click **OK**. The style appears below the currently active style in the Paragraph Styles panel. If you enter a shortcut, this appears to the right of the style name.

5 A useful alternative technique for creating a paragraph style is to base the style on existing formatting already applied to a paragraph. Start by creating character and paragraph formatting on a paragraph that defines the way you want your text to appear. Make sure the Text Insertion Point remains located in the paragraph, then choose **New Paragraph Style** from the panel menu. The **New Paragraph Styles** dialog box picks up the settings from the selected text – these are listed in the **Style Settings** box. If necessary, click on the Formatting categories on the left of the dialog box and create

or adjust settings as required. Click **OK** when you are satisfied with your settings.

When you import a Word document, you may find that **Word styles** are also imported. A disk drive icon (📥) appears in the Paragraph Styles panel to indicate a style imported from Word:

To prevent InDesign from importing Word styles when you place the Word file, select the **Show Import Options checkbox** in the Place dialog box. When you click **Open** in the **Microsoft Word import options** dialog box, select the **Remove Styles and Formatting from Text and Tables** checkbox.

135

When you create a paragraph style using the technique in Step 5, select the **Apply Style to Selection** checkbox to apply the style to the text on which it is based:

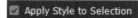

Character Styles

Use the technique detailed in Step 5 on page 135 to create a character style based on existing formatting.

Character styles are used at a sub-paragraph level and control only the character attributes of type – they do not affect paragraph-level attributes such as alignment and indents. Character styles are used to control the appearance of anything from a single character to a word, phrase, sentence, or group of sentences.

As with paragraph styles, character styles enable you to style text consistently and efficiently throughout a document and across publications.

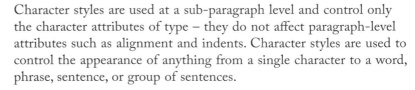

1 To create a character style, choose **Type** > **Character Styles** (**Shift + F11**) to show the Character Styles panel, or click the **Character Styles** icon if the panel is docked in the Panel Dock. There are no existing character styles to choose from in the default Character Styles panel.

You can only use numbers on the number keypad as shortcuts for paragraph and character styles – you cannot use letters or non-keypad numbers.

136

2 Choose **New Character Style...** from the Character Styles panel menu (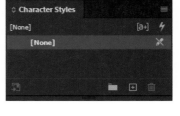).

3 Enter a name for the style in the **Style Name** entry field.

In the Essentials workspace, in the **Text Style pane** of the Properties panel, click the **Paragraph Style** button () or the **Character Style** button (**A.**) to display the Paragraph and Character Styles panels respectively.

④ Make sure **Based On** is set to **[None]** to simplify working with styles when you first start to use them.

⑤ Enter a shortcut for the character style in the Shortcut field if desired. Shortcuts allow you to apply character styles using the keyboard instead of the mouse, and can be very useful. Make sure **Num Lock** is **On** to set a shortcut, then hold down any combination of **Ctrl/Command**, **Alt/Option** and **Shift**, and type a number on the numeric keypad.

⑥ Click on style options in the formatting categories pane on the left for which you want to define settings. (Individual options available are covered in Chapter 4).

General
Basic Character Formats
Advanced Character Formats
Character Color
OpenType Features
Underline Options
Strikethrough Options
Export Tagging

⑦ Choose **Character Color** from the formatting categories pane if you want to specify a different fill and/or stroke color for the text. Make sure you select either the **Fill** or **Stroke** icon (as appropriate) before you click on a color swatch.

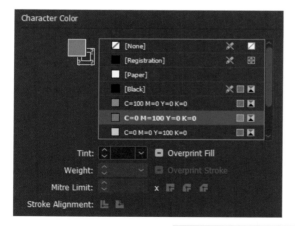

⑧ **OK** the dialog box. The character style, together with any keyboard shortcut, appears in the panel. (See pages 138-139 for details on applying character styles.)

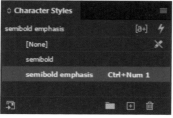

In the Paragraph Styles and Character Styles panels you can use the **New Style Group command**, in the panel menu, to order and group similar styles into a folder. After you create a new style group, drag styles onto the group folder to move them into an appropriate category:

Click the **Expand/Collapse** button (▶) to reveal/hide the contents of style group folders.

One of the most significant advantages of using paragraph and character styles is that if you edit the style description, all text to which the style has been applied updates automatically.

Applying and Editing Styles

Before you apply paragraph or character styles, make sure you highlight an appropriate range of text.

1. To apply a paragraph style, select the **Type tool**; click into a paragraph to apply the style to one paragraph only, or highlight a range of paragraphs.

2. Click on the paragraph style name in the Paragraph Styles panel to apply it. Alternatively, make sure that **Num Lock** is **On**, and enter the shortcut you specified when you set up the style. Character styles and any local formatting overrides are retained when you apply a paragraph style.

3. To remove all current character styles and local formatting as you apply a paragraph style, hold down **Alt/Option + Shift**, and click the paragraph style name in the Paragraph Styles panel.

4. To preserve all character styles in the text but remove local formatting overrides as you apply a paragraph style, hold down **Alt/Option**, and click on the paragraph style name.

5. **[Basic Paragraph]** is a default style that is present in each new document you create and is initially applied to text you type. Double-click the **[Basic Paragraph]** entry in the Paragraph Styles panel to edit it. You cannot delete or rename this style.

When your Text Insertion Point is in a paragraph of text, the top row of the Paragraph Styles panel displays the name of any paragraph style applied to the paragraph:

A "+" symbol appears next to a style name if, in addition to a paragraph or character style, the selected text has a "**local**" or manual formatting override applied to it:

A local or manual formatting "**override**" refers to formatting that you have applied by highlighting a range of text and using controls in the Character, Paragraph and Control panels, or the Type menu, to create settings in addition to the paragraph or character style already applied.

138

Oborpero stionse quamconse tat nibh et, **character style applied** euis nos aliquatinit, si. Feum adiamcore *local formatting applied* delessed tat. Ut aci tat. Uptat local formatting applied ute commy nulluptat. Quis ad et **character style** odolobor sed tis *local formatting appli* faccum zzril ing esse volor ad er El dit at incilis inis ea commy onulputatio odo

Oborpero stionse quamconse tat nibh et, **character style applied** euis nos aliquatinit, si. Feum adiamcore *local formatting applied* delessed tat. Ut aci tat. Uptat local formatting applied ute commy nulluptat. Quis ad et atue tem iusto od **character style applied** dipsumm odolobor sed tis aci tat et *local formatting applied* elendionse ea faccum zzril ing ex exerci blandion esse volor ad er in dolor si. El dit at incilis diametue conum inis ea commy nonullam, veniamc onulputatio odolore incidunt essi.

Oborpero stionse quamconse tat nibh et, character style applied euis nos aliquatinit, si. Feum adiamcore local formatting applied delessed tat. Ut aci tat. Uptat local formatting applied ute commy nulluptat. Quis ad et atue tem iusto od character style applied dipsumm odolobor sed tis aci tat et local formatting applied elendionse ea faccum esse vc El dit a inis ea onulpu

Oborpero stionse quamconse tat nibh et, **character style applied** euis nos aliquatinit, si. Feum adiamcore local formatting applied delessed tat. Ut aci tat. Uptat local formatting applied ute commy nulluptat. Quis ad et atue tem iusto od **character style applied** dipsumm odolobor sed tis aci tat et local formatting applied elendionse ea faccum zzril ing ex exerci blandion esse volor ad er in dolor si. El dit at incilis diametue conum inis ea commy nonullam, veniamc onulputatio odolore incidunt essi.

6 To apply a character style, highlight a range of characters – from a single character upward. Click on a character style in the Character Styles panel. You can also use a shortcut if you set one up, but make sure that **Num Lock** is **On** before you use a keyboard shortcut to apply a style. Character styles change only the settings that are specified in the style.

> Oborpero stionse quamconse tat nibh et, bold brown character style applied euis nos aliquatinit, si. Feum adiamcore local formatting applied delessed tat. Ut aci tat. atting applied

> Oborpero stionse quamconse tat nibh et, **bold brown character style applied** euis nos aliquatinit, si. Feum adiamcore local formatting applied delessed tat. Ut aci tat. Uptat local formatting applied

You can also clear all formatting overrides in a selection by clicking the **Clear Formatting Overrides button** in the bottom of the Paragraph Styles panel:

Hold down **Ctrl/Command** then click the button to selectively clear **only character formatting overrides**.

Hold down **Ctrl/Command + Shift** then click the button to clear **paragraph formatting overrides only**.

Editing Styles

The basic principles for editing paragraph and character styles are the same. When you edit a style, text to which the style has been applied is updated automatically to reflect the new settings. This is a powerful reason for working with paragraph and character styles.

1 To edit a style, click on the style you want to edit in either the Paragraph or Character Styles panel. Choose **Style Options** from the panel menu (▤).

Select the Preview option in the **Style Options** dialog box to see a preview of the result of changes you make to settings before you **OK** the dialog box.

2 Modify settings in the **Style Options** dialog box. The options are identical to those available when you first set up the style. Click **OK**. The change is applied throughout the document wherever the style is already applied.

3 Use the **Redefine Style** command in the panel menu to redefine a style based on selected text. First, select some text currently formatted with the style you want to redefine. Change **Character** and **Paragraph** settings as required, then choose **Redefine Style** from the panel menu.

4 To delete a style, click on the style name to select it then choose **Delete Style** from the panel menu, or click the **Trash** icon at the bottom of the panel. Use the warning box to replace the deleted style, wherever it is used in the document, with another available style.

Click the **Reset to Base button** to quickly reset a child style to match the parent style on which it is based.

Copying Styles

To save the work of recreating a complex set of both paragraph and character styles when you need to establish a consistent identity across a range of publications, you can copy individual styles or complete sets of styles from one document to another.

Use the keyboard shortcuts **F11** to show the Paragraph Styles panel and **Shift + F11** to show the Character Styles panel, if they are not already showing.

1 To load styles into a document, first make sure the Paragraph or Character Styles panel is showing. From the panel menu (■), choose **Load Paragraph/Character Styles...**, depending on which panel is active.

Load Paragraph Styles...
Load All Text Styles...

2 In the **Open File** dialog box, use standard Windows/Mac techniques to navigate to the document with the styles you want to copy. Click on the name of the file, then click the **Open** button. Use the checkboxes on the left of the **Load Styles** dialog box to specify the styles you want to import. You can click the **Check All** or **Uncheck All** buttons to quickly select or deselect all available styles. Click **OK** when you finish making your selection, to copy the styles into your InDesign document.

To copy all paragraph and character styles at the same time, choose **Load All Text Styles** from the panel menu.

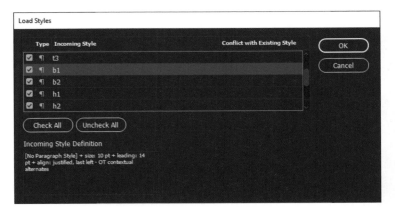

3 An alternative technique is to highlight some styled text in a different document. Choose **Edit > Copy** to copy the text to the Clipboard. Move to the document into which you want to copy the styles. Create a text frame, then paste the text into the frame. The text is pasted into the document and any paragraph or character styles are appended to the document's Paragraph and Character Styles panels. This is a useful technique for selectively copying styles from one document to another.

When you create a new paragraph or character style, you can add it to a CC library if required. In the New Paragraph/ Character Style dialog box, click the **Add to CC Library** checkbox, then from the library pop-up menu, select the CC library where you want to add the style.

12 Tables and Tabs

Whenever you need to organize data into regular rows and columns, the table functionality built into InDesign offers a wide variety of flexible and creative options. Use tabs to align columns of figures or text accurately in tabular format.

Inserting a Table

Typically, a table consists of a group of cells, arranged in a grid-like structure, that hold related data, which can consist of text and/or images. In many instances, you have greater control using a table than using tabs to create the regular horizontal and vertical spacing of information across and down the page, especially when the graphic appearance of the table is an important consideration.

You can insert a table at the Text Insertion Point in the middle of running copy. The table can then be handled like an inline graphic. You can use the **Space Before/Space After** fields in the **Table Setup** dialog box (**Table > Table Options > Table Setup**) to control the amount of space above and below the table.

1 Use the **Type tool** to draw a text frame that is roughly the size of the table you want to create. You do not have to be exact at this stage, as you can make the frame fit the final size of the table later.

2 With the Text Insertion Point located in the frame, choose **Table > Insert Table**. In the **Insert Table** dialog box, enter the number of columns and rows you want in the table. Enter values for **Header Rows** and **Footer Rows** if your table will span across columns or pages and you want certain rows to repeat at the top (Header Rows) and/or bottom (Footer Rows) of each column/page the table covers.

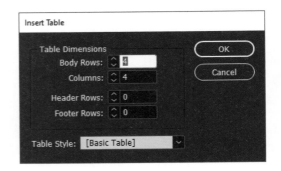

3 Select a style for the table from the **Table Style** pop-up menu, if you have previously created one.

4 Click **OK**. The table structure appears in the text frame. InDesign creates a table with default-size cells that

fill the width of the text frame. The height of the cells is initially determined by the slug height of the type, which is relative to the default type size that is currently set. The table has 1-point vertical and horizontal grid lines applied by default.

In the context of table cell height, the "slug" height refers to the size of the black highlighted area of selected text:

Adding and Deleting Rows/Columns

Unless you know the exact contents of a table at the outset, you will probably need to add or delete rows and/or columns to achieve the final table structure you require.

1 To change the number of rows/columns in a table, make sure you are working with the **Type tool**. Click into the table to place the Text Insertion Point in a cell.

2 Choose **Table** > **Table Options** > **Table Setup** (**Ctrl/Command** + **Alt/ Option** + **Shift** + **B**). Enter new values for the **Body Rows** and **Columns** fields as required. **OK** the dialog box. If you reduce the number of rows or columns, a warning prompt appears. Click **OK** if you want to proceed.

3 Alternatively, use the Table panel to change the number of rows/columns. Choose **Window** > **Type & Tables** > **Table** (**Shift** + **F9**) to show the Table panel if it is not already showing. Change the values in the **Number of Rows/Columns** fields as required.

When you use the **Table Options dialog box** or the **Table panel** to change the number of rows or columns, rows are added or deleted along the bottom of the table, and columns are added or deleted along the right-hand edge of the table.

4 For more precision when adding rows or columns, position your cursor in a cell, and then choose **Table** > **Insert** > **Column** or **Row**. Enter the number of columns or rows you want to create, and specify whether you want the columns inserted to the right or left of the column where the cursor is located, or rows above or below where the cursor is located.

Get into the habit of choosing the **Type tool** when you work with tables. Most actions you perform on a table use this tool.

Highlighting and Moving Techniques

To select an individual cell, position the Text Insertion Point in the cell, and then choose **Table > Select > Cell** (Ctrl/ Command + /).

Highlighting

To control the formatting and appearance of a table, you must be able to highlight its various parts to suit the task in hand. The most essential tool for working with tables is the **Type tool**.

① To select the entire table, working with the **Type tool**, position your cursor on the top-left corner of the table. Click once when the cursor changes to the **Table select cursor** (⭦). This is useful when you want to work globally on the table – for example, to set type size and font for every cell in the table.

Schnobler	1 Year Bond: maturity date 12/11/23	2 Year Bond: maturity date 12/11/24	3 Year Bond: maturity date 12/11/25	10 Year Bond: maturity date 12/11/33
Issue No.	ZX	ZY	ZZ	ZZZ
Year 1	2.35%	2.55%	2.85%	2.95%
Year 2	2.95%	2.35%	2.35%	2.65%
Year 3	2.95%	2.35%	2.35%	2.65%
Year 4	3.95%	3.95%	3.95%	3.95%

You can also highlight the table, columns and rows using the **Table > Select** sub-menu or keyboard shortcuts:

Cell	Ctrl+/
Row	Ctrl+3
Column	Ctrl+Alt+3
Table	Ctrl+Alt+A
Header Rows	
Body Rows	
Footer Rows	

② To select an entire row, position your cursor on the left edge of the row you want to select. Click when the cursor changes to the **Row select cursor** (→).

③ To select an entire column, position your cursor on the top edge of the column you want to select. Click once when the cursor changes to the **Column select cursor** (↓).

Moving from Cell to Cell

① To move the Text Insertion Point from cell to cell, press the **Up/ Down/Left/Right** arrow keys. You can also press the **Tab** key to move the cursor cell by cell to the right. To move the Text Insertion Point backward through the table, hold down **Shift** and press the **Tab** key.

If you press the **Tab key** when the Text Insertion Point is located in the last cell of the table, you create an additional row in the table.

② In large tables with many rows, it can be useful to jump quickly to a specific row. Choose **Table > Go to Row**. Enter the row

Go to Row		
Body	47	OK
		Cancel

number and then click **OK**. Use the **Row** pop-up menu to choose Header or Footer rows if required.

Resizing Columns and Rows

As you create a table, you will need to control the width of columns and the height of cells to create the table structure you require. You can resize columns and rows manually with the mouse, or you can enter exact values to achieve the results you want. Remember to work with the **Type tool** when you want to make changes to the structural appearance of a table.

The slug height (see page 142) determines the minimum height for cells in the row, even if there is no actual type in any of the cells. When you drag a cell border to resize a row's height, you cannot make it shorter than the height needed to accommodate the slug. Reduce the type size set for a row if necessary.

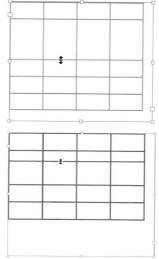

1   To resize the height of an individual row, position your cursor on the row border, then drag up or down. This changes the height of the row, and the height of the table also adjusts accordingly. Hold down **Shift** and drag a **row border** to restrict the resizing to the two rows that share the border you drag – one row gets bigger, the other smaller, but the overall size of the table remains unchanged.

2   To resize all rows in the table proportionally, position your cursor on the bottom edge of a table, hold down **Shift**, then **drag**.

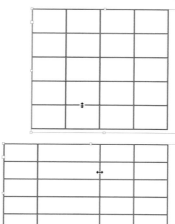

3   To resize the width of an individual column, position your cursor on the column border, and drag left or right. This changes the width of the column, and the width of the table also adjusts accordingly. Hold down **Shift** and drag a **column border** to restrict the resizing to the two columns that share the border you drag – one column gets wider, the other narrower, but the overall width of the table remains unchanged.

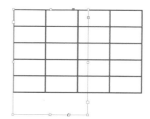

As you make changes to the columns and rows in a table, the table itself may extend beyond the boundaries of the text frame in which it is located:

In this case, you can either use the **Selection tool** to change the width of the text frame manually or choose **Object > Fitting > Fit Frame to Content** to match the size of the frame to the size of the table it contains.

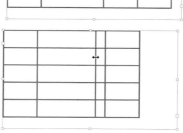

...cont'd

Hold down **Alt/Option** and press **Page Up/Page Down** to move the cursor to the first/last cell in the column. Hold down **Alt/Option** and press **Home/End** to move the cursor to the first/last cell in a row.

4. To resize all columns proportionally, position your cursor on the right edge of a table, hold down **Shift**, and then **drag**.

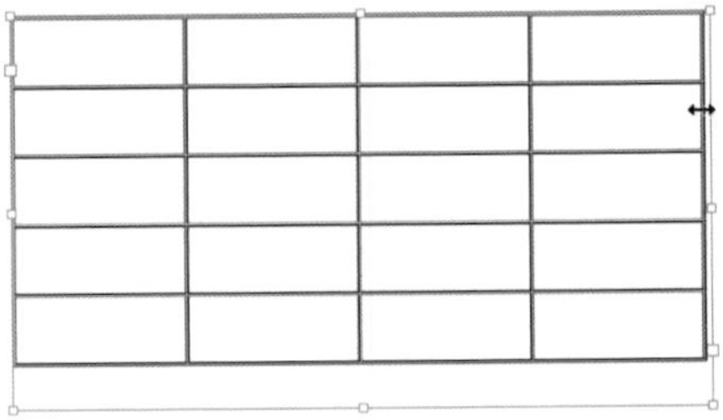

Resizing Columns and Rows Precisely

1. To resize a row to an exact height, select the row using the **Row selection cursor** (➜). (Drag with the **Row selection cursor** to select more than one consecutive row.) Choose **Table** > **Cell Options** > **Rows and Columns**. With the **Rows and Columns** tab selected, select **Exactly** from the **Row Height** pop-up, then enter the row height you require in the entry field.

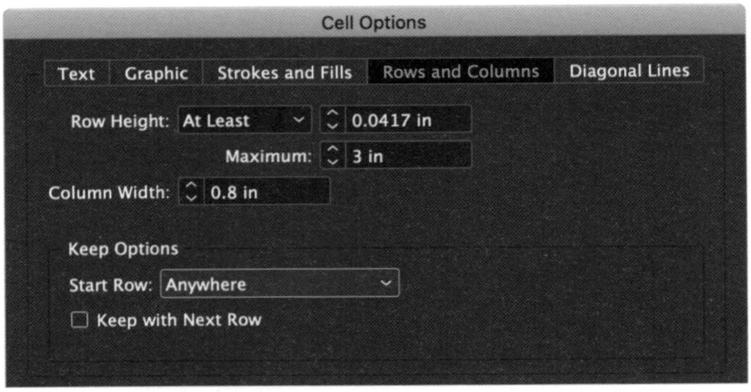

Choose **Window** > **Type & Tables** (**Shift** + **F9**) to show the Table panel if it is not already showing. Remember to rest your cursor on the icons in the panel to reveal the **Tool Tips**, which will help you identify the control you want to use:

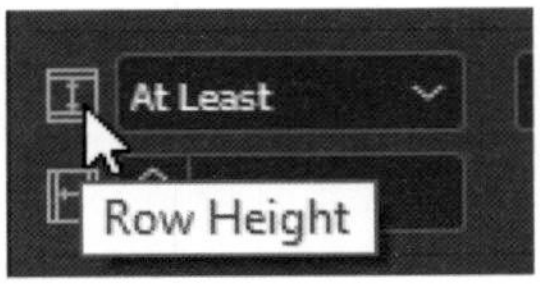

2. To resize a column to an exact width, select the column using the Column selection cursor (🡇). (Drag with the Column selection cursor to select more than one consecutive column.) Choose **Table** > **Cell Options** > **Rows and Columns**. Enter the width you require in the **Column Width** entry field.

3. As an alternative to using the **Cell Options** dialog box, you can use the Table panel to specify exact width and height settings for selected rows or columns.

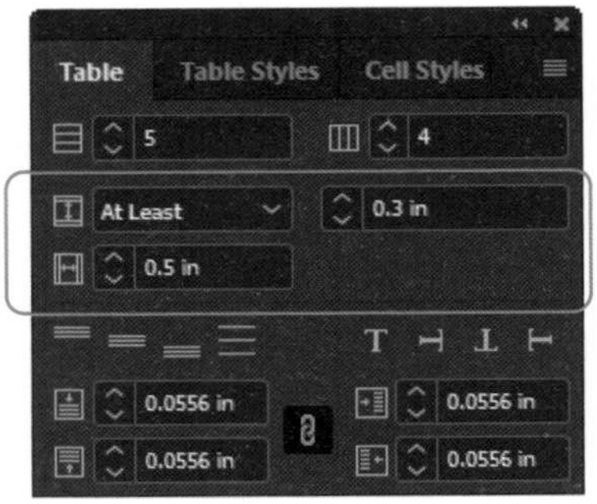

Entering Content in Cells

When you have a suitable table structure, you can start to enter text and images into individual cells.

Entering Text

1 Select the **Type tool**, then click into a cell to place the Text Insertion Point. Begin typing. Text will wrap when it reaches the edge of the cell. Provided that you have not specified an exact height for the row, the cell will expand downward as you enter more and more text.

2 If you have set an exact height for the row and you enter more text than will fit, the additional text becomes overset text and the overset cell marker appears ( ).

3 To highlight overset text in a cell (for example, to reduce the type size so that it fits into the cell), click to place the Text Insertion Point in the cell, and then choose **Edit** > **Select All** (**Ctrl/Command + A**).

Placing an Image in a Cell

1 To place an image into a table cell, it is best if you make sure that the image will fit into the cell before you place it. Working with the **Type tool**, click into the cell to place the Text Insertion Point. Choose **File** > **Place**, and then use standard Windows/Mac techniques to locate the file. Select the file, and click the **Open** button to place the image into the cell.

2 Depending on the relative size of the image/graphic and the size of the cell, the image fills the cell proportionally. In this case, the complete image may not be visible. To manipulate the image so that it fits into the cell, select it using the **Selection tool**. Use the techniques covered in Chapter 6 to achieve the result you require.

If you have specified an exact height for a row and you enter more text than will fit into the cell, the excess text becomes **overmatter**. You cannot thread overset text into another cell.

You can also paste text from the Clipboard into a cell, or use **File** > **Place** to import text into a cell.

For an image pasted into a cell that extends beyond the right-hand edge of the cell, position your text cursor in the cell, then select the content of the cell by pressing the **Esc** key. Choose **Table** > **Cell Options** > **Text**.

Select the **Clip Contents to Cell checkbox** to hide any part of the image that extends beyond the right-hand edge of the cell:

Cell Controls – Text

Use the **Text** tab of the **Cell Options** dialog box to control how text sits within a cell. The range of available options is similar to those found in the **Text Frame Options** dialog box.

 Deselect the **Make all settings the same button** ( ) to set individual top, bottom, left and right insets.

1 Working with the **Type tool**, make sure you have the Text Insertion Point located in a cell, or select columns, rows or an entire table, depending on the range of cells you want to change. Choose **Table** > **Cell Options** > **Text**.

2 To create additional space on the inside of a cell, enter Cell Inset values for Top/Bottom/Left/ Right as required.

Schnobler	1 Year Bond maturity date 10/11/23	2 Yea matu 10/1
Issue No.	ZX	ZY

 Select the Preview option in the **Cell Options dialog box** to see the effect of values you enter before you **OK** the dialog box.

3 To align content vertically in a cell (for example, if you want to center text in a cell vertically), select an option from the **Vertical Justification Align** pop-up menu.

Schnobler	1 Year Bond maturity date 10/11/23
Issue No.	ZX

Controls for insets, vertical alignment and rotation are also available in the bottom portion of the Table panel:

4 To rotate text in a cell, choose a rotation amount from the **Rotation** pop-up menu. The overset cell marker (⊣) appears if the rotated text does not fit within the dimensions of the cell.

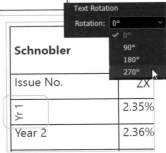

Schnobler	
Issue No.	ZX
Yr 1	2.35%
Year 2	2.36%

Cell Strokes, Borders and Fills

When you create a table, it appears with a default 1-point black border and 1-point black vertical and horizontal grid lines. You can hide or show the border and grid lines, and there is a wide variety of customization options.

Make sure you select a line type when you create settings for a border. If you set a weight but the **Type** field is set to **None**, the border does not appear.

1 To make changes to the border, select the **Type tool**, and then click into a cell in the table. Choose **Table** > **Table Options** > **Table Setup**. Use the **Table Border** pop-up menu to control the appearance of the border. To remove the border, either set the **Weight** field to zero or select **None** from the **Type** pop-up menu. Click **OK** to apply the settings.

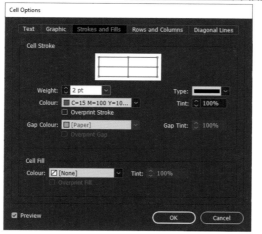

2 To make changes to all vertical and horizontal grid lines as well as the border, select the entire table using the Table select cursor, and then choose **Table** > **Cell Options** > **Strokes and Fills**. Make sure that all the Preview proxy lines are set to blue. Use the **Cell Stroke** pop-up menu to set the appearance of the lines for the table. Click **OK**.

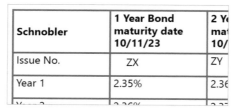

In the **Strokes and Fills** tab of the **Cell Options** dialog box, click the proxy lines to toggle them from **blue (selected)** to **gray (not selected)**. Changes to settings are applied to cell borders represented by the blue lines, and do not affect cell borders represented by the gray lines:

149

...cont'd

3 To make changes to an individual cell or a series of

selected cells, either click into a single cell or highlight a range of cells. Choose **Table > Cell Options > Strokes and Fills**. In the **Cell Stroke** tab of the **Cell Options** dialog box, click the proxy lines

Schnobler	1 Year Bond maturity date 10/11/23	2 Year Bond maturity date 10/11/24	3 Year matur 10/11/
Issue No.	ZX	ZY	ZZ

to toggle them from blue (selected) to gray (not selected) so that the settings you create are only applied to the selected borders for the selected cells.

Cell Fill

As well as controlling stroke attributes for cell borders in a table, you can specify the fill color for individual cells, a range of highlighted cells, or the entire table.

1 Make sure you are working with the **Type tool**, and select a range of cells you want to change.

2 Choose **Table > Cell Options > Strokes and Fills**. In the **Cell Fill** area of the dialog box, use the **Color** pop-up menu to select a color from the existing range of color swatches available in the Swatches panel. Use the **Tint** entry box to specify a tint from 0-100%, if required. **OK** the dialog box to apply the setting.

Schnobler	1 Year Bond maturity date 10/11/23	2 Year Bond maturity date 10/11/24	3 Ye mat 10/1
Issue No.	ZX	ZY	ZZ

Working with the **Advanced** or **Essentials Classic** workspace selected, with a cell, row or column selected you can access controls for stroke and fill in the Control panel along the top of the InDesign workspace:

You can also use the Swatches panel to apply fill color to selected cells.

Alternating Fills and Strokes

InDesign tables provide a range of versatile controls for setting up alternating fill patterns for rows and columns, and also alternating stroke controls for vertical and horizontal grid lines in a table. Using alternating fill and/or stroke controls, especially in a complex table with a considerable number of rows, can help make the data it contains be more readable and more easily understood.

Alternating fill controls apply to an entire table. Select the **Preserve Local Formatting option** in the Table Options dialog box if you want to retain formatting already applied to specific cells, rows or columns.

Alternating Fills

① To set up alternating fills for a table, working with the **Type tool**, click in a cell to position the Text Insertion Point. Choose **Table > Table Options > Alternating Fills**.

② Select an option from the **Alternating Pattern** pop-up to specify the frequency with which the pattern repeats. You can also set up custom patterns by entering values in the **First** and **Next** entry boxes: enter values that are different in **First** and **Next** to set up an irregular repeating pattern.

You can specify alternating patterns for rows or columns in a table, but not both. Use the same principles for alternating columns as those demonstrated on this page for alternating rows.

③ Use the **Color** pop-ups to select colors from the existing range of colors in the Swatches panel. Specify a tint using the **Tint** entry fields.

Schnobler	1 Year Bond maturity date 10/11/23	2 Year Bond maturity date 10/11/24	3 Year Bond maturity date 10/11/25	10 Year Bond maturity date 10/11/33
Issue No.	ZX	ZY	ZZ	ZZZ
Year 1	2.35%	2.36%	2.37%	2.96%
Year 2	2.36%	2.37%	2.39%	3.33%
Year 3	2.42%	2.48%	2.56%	3.98%
Year 4	2.61%	2.53%	2.59%	4.02%

④ Enter values for **Skip First/Skip Last** for rows at the top or bottom of the table that you do not want included in the repeating pattern because you want to format them individually.

Skip First: 2 Rows   Skip Last: 0 Rows

Schnobler	1 Year Bond maturity date 10/11/23	2 Year Bond maturity date 10/11/24	3 Year Bond maturity date 10/11/25	10 Year Bond maturity date 10/11/33
Issue No.	ZX	ZY	ZZ	ZZZ
Year 1	2.35%	2.36%	2.37%	2.96%
Year 2	2.36%	2.37%	2.39%	3.33%
Year 3	2.42%	2.48%	2.56%	3.98%
Year 4	2.61%	2.53%	2.59%	4.02%

...cont'd

5 If you have formatted some cells individually, you can retain the individual formatting characteristics when using alternating fill controls by selecting the **Preserve local formatting** option in the **Fills** tab of the **Table Options** dialog box.

Alternating Strokes

Setting up alternating stroke patterns on the vertical and/or horizontal grid lines of a table is similar to setting up alternating fill patterns. You can combine alternating fills and alternating strokes.

1 To set up alternating strokes for a table, working with the **Type tool**, click in any cell to position the Text Insertion Point. Choose **Table > Table Options > Alternating Row Strokes**.

2 Select an option from the **Alternating Pattern** pop-up to specify the frequency with which the alternating stroke pattern repeats. You can also set up custom patterns by entering values in the **First** and **Next** entry boxes: enter values that are different in **First** and **Next** to set up an irregular repeating pattern.

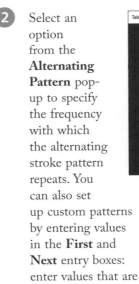

The **Column Strokes tab** of the Table Options dialog box provides exactly the same set of options as the **Row Strokes** tab.

3 Use the **Color** pop-ups to select stroke colors from the swatches available in the Swatches panel. Specify a tint using the **Tint** entry fields.

4 Enter values for **Skip First/Skip Last** for rows at the top or bottom of the table that you do not want included in the repeating pattern.

Importing Tables

You can import Excel data and Word table data either as an
InDesign table or as tabbed text, to suit your requirements.

If you have
imported
tabbed text
from another
application, you
can convert it into a table.
Select the tabbed text
using the **Type tool**, then
choose **Table** > **Convert
Text to Table** to convert
the text into an InDesign
table with default
formatting.

1 With the Text Insertion Point located in the text frame, choose
File > **Place**. Use standard Windows/Mac techniques to navigate
to the file you
want to import.
Click on the
file to select it,
select the **Show
Import Options**
checkbox, then
click the **Open**
button.

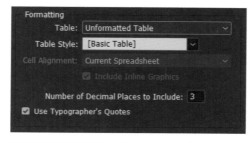

2 For Excel
spreadsheet
data, use the
**Formatting:
Table** pop-up
menu to specify
whether you
want a **Formatted Table** (retaining formatting from Excel), an
Unformatted Table or **Unformatted Tabbed Text**.

Schno-bler	1 Year Bond maturity date 10/11/23	2 Year Bond maturity date 10/11/24	3 Year Bond maturity date 10/11/25	10 Year Bond maturity date 10/11/33
Issue No.	ZX	ZY	ZZ	ZZZ
Year 1	2.35%	2.36%	2.37%	2.96%
Year 2	2.36%	2.37%	2.39%	3.33%
Year 3	2.42%	2.48%	2.56%	3.98%
Year 4	2.61%	2.53%	2.59%	4.02%

3 For Word table
data, use the
**Formatting:
Convert Table
To** pop-
up menu to
control whether
imported data
appears as an
**Unformatted
Table** or
**Unformatted
Tabbed Text**.
Alternatively,
you can click
the **Preserve Styles and Formatting** radio button for InDesign to
preserve as much of the original Word formatting as possible.

Formatting
- Remove Styles and Formatting from Text and Tables
- ☐ Preserve Local Overrides
- Convert Tables To: Unformatted Tabbed Text
- ○ Preserve Styles and Formatting from Text and Tables
- Manual Page Breaks: Preserve Page Breaks

Schnobler — 1 Year Bond maturity date 10/11/23 — 2 Year Bond maturity date 10/11/24 — 3 Year Bond maturity date 10/11/25 — 10 Year Bond maturity date 10/11/33
Issue No. — ZX — ZY — ZZ — ZZZ
Year 1 — 2.35% — 2.36% — 2.37% — 2.96%
Year 2 — 2.36% — 2.37% — 2.39% — 3.33%
Year 3 — 2.42% — 2.48% — 2.56% — 3.98%
Year 4 — 2.61% — 2.53% — 2.59% — 4.02%

When you
paste Excel or
Word table
data into
InDesign from
the Clipboard, the
**Clipboard Handling
Preferences setting**
determines how the data
appears in InDesign:

When Pasting Text and Tables from Other Applications
Paste:
○ All Information (Index Markers, Swatches, Styles, etc.)
● Text Only

With **Text Only** selected,
imported data appears
as tabbed text. With **All
Information** selected, it
appears as an unformatted
InDesign table.

Setting and Editing Tabs

Use tabs to line up columns of figures and text accurately. There is a temptation to use spaces for aligning entries in tabular information, but this can be inaccurate and unnecessarily time-consuming. With practice, you will come to appreciate the accuracy and versatility of tabs for producing professional results.

Tabs are a paragraph attribute. If your Text Insertion Point is flashing in a paragraph when you enter the **Tabs** panel, you set tabs for that specific paragraph. Remember to highlight a range of paragraphs to set or edit tabs for more than one paragraph.

When you begin working with tabs it is often easiest to set the tabs when the text frame is empty, and then enter your text and adjust the tabs as necessary to get the table to work.

1 To set tabs for an empty text frame, click into it with the **Type tool**, to place the Text Insertion Point. Choose **Type > Tabs**. The **Tabs** panel appears along the top of the selected frame. Provided the top of the frame is visible, the panel snaps to the top of the frame and matches the width of the first column. Initially, the left edge of the frame and the zero point of the **Tabs** panel ruler line up. This is useful as a visual reference for setting tabs. If you reposition the **Tabs** panel and then want to realign it to the text frame, or if you want to snap the **Tabs** panel to the same width as the text frame, click the **Position Panel above Text Frame** button (🔲) to the right of the panel.

The **gray tick marks**, set at regular intervals, that appear initially on the Tab ruler represent the **default tab stops**.

As you set your own tab stops, default tabs to the left of the tab you set disappear automatically.

2 Choose a tab alignment type by clicking on one of the tab alignment buttons: **Left**, **Center**, **Right**, or **Decimal**.

If necessary, you can manually reposition the **Tabs panel** so that the zero point of the ruler lines up with the left edge of the text frame, and then use the resize icon of the panel to match the width of the ruler to the width of the text frame.

3 Click on the tab ruler to position a tab manually. The tab is indicated by a tab marker on the ruler. Initially, the tab is selected, indicated by a blue highlight on the tab (🔳). Drag the tab to fine-tune its position. Alternatively, enter a value in the **X** entry field to specify the position for the selected tab. Press **Enter/Return** to set the tab marker on the ruler.

...cont'd

④ Repeat Steps 2-3 on the previous page until you have set as many tabs as you need. Either close the panel ( ), or leave it showing until you have finished fine-tuning the table.

⑤ Enter text for the table in the text frame. Press the **Tab** key each time you need to line up text or numbers at a particular tab-stop position.

To set tabs at equal distances across the text frame, set the first tab marker, and then choose **Repeat Tab** from the Tabs panel menu ( ):

| Clear All |
| Delete Tab |
| Repeat Tab |
| Reset Indents |

Deleting Tabs

You can delete individual tabs or all tabs for a paragraph or a range of paragraphs. Deleting all tabs can be useful when you receive text that has been set up with tabs in a word processing application. Typically, such tabs are not going to work in a page layout with different margin and column settings.

The text frame remains active when the Tabs panel is displayed above it. This means you can edit and adjust text as you create and adjust tabs.

① With the **Type tool**, click into a paragraph in a text frame, or highlight a range of paragraphs. Choose **Type** > **Tabs**. If you highlight a range of paragraphs containing different tab settings, the tab markers for the first highlighted paragraph appear as normal; tab markers for the other highlighted paragraphs with different tab stops appear gray. It is best to avoid deleting tabs when you have mixed tab settings. Reselect paragraphs more accurately to avoid getting mixed settings.

② Drag the tab marker you want to delete off the Tabs panel ruler. Do not press the **Backspace** or **Delete** keys – these will delete highlighted text and not tab markers. To delete all tabs, choose **Clear All** from the Tabs panel menu.

Editing Tabs

Getting a tabbed table to work can sometimes be a tricky business. You often end up having to edit tabs and fine-tune text to get a polished result. When editing tabs, take care to select the specific range of paragraphs whose tabs you want to change.

The most common problem you encounter initially when you are learning to work with tabs is setting type sizes that are too large, so that entries don't fit between the tab stops.

Either reduce the size of your type or enter less text between the tab stops.

...cont'd

1 To edit tabs, click once on a tab marker to highlight it. Its numeric position appears in the **X** entry field. Enter a new value, and then press **Enter/Return** to apply the change. The tab marker moves to its new position and text in the highlighted paragraphs moves according to the new tab setting.

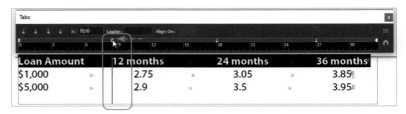

2 You can manually adjust a tab marker by dragging it along the tab ruler. As you do so, a vertical line appears to indicate the position of the tab in the table. This can be extremely useful as an aid in positioning the tab marker exactly where you want it. The **X** entry field gives a numeric read-out as you drag the tab marker. Text in the highlighted paragraphs realigns when you release the mouse. Repeat the process as necessary, then close the Tabs panel when you are satisfied.

Leader Dots

You can set a tab stop to have leader dots extending to the tab position. Leader dots help the eye to follow across a row of information in detailed tables, such as railway timetables, which consist of columns of figures that are not easily differentiated.

1 To set leader dots for a tab, in the tab ruler click on the tab that you want leader dots to run up to. The tab is highlighted and its position appears in the **X** field.

2 Enter up to eight characters in the **Leader** entry field. You will typically use a full stop. You can create a less dense leader by entering a combination of full stops and spaces.

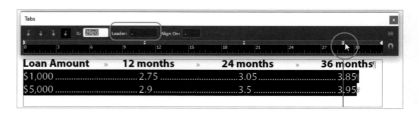

13 Printing and Exporting

This chapter shows you how to print proof copies of your documents to an inkjet or laser printer, how to create a package to transfer projects effectively, and how to export documents as PDF files.

Printing a Composite

When you print a composite, all colors or shades in the document are printed on one sheet of paper. Printing from InDesign follows standard Windows and Mac principles.

InDesign supports printing to both PostScript® and non-PostScript printers (such as low-cost color inkjet printers).

Print settings you create in the Print dialog box are saved with the document.

In a document with mixed page sizes, click the **Select matching page sizes** button to select all pages of the same size as the current page. You cannot print pages of different sizes at the same time:

Click the **Save Preset** button to save the current print settings as a preset, which you can then select from the **Print Preset** pop-up when you need to reuse the same settings.

1 To print a composite proof, choose **File > Print**. Select your printer from the **Printer** pop-up menu. Enter the number of copies you want to print. In the **Pages** area, make sure the **All Pages** radio button is selected

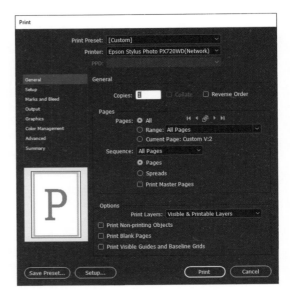

to print all pages in the document, or click the **Range** button to print specific pages. Enter the page numbers you want to print in the **Range** entry field. To specify a continuous range of pages, enter numbers separated by a hyphen – e.g. 10-15. To specify individual pages, enter numbers separated by a comma – e.g. 3, 6, 12. You can combine both techniques – e.g. 1-4, 8, 10, 12-15.

2 Select checkboxes in the **Options** area as required if you want to print page elements that do not normally print:

non-printing objects, blank pages or non-printing guides. Use the **Print Layers** pop-up to control which layers print.

3 Click printing categories in the Categories list to create settings for each in turn. Select the **Setup** category to set options for **Paper Size**, **Orientation** and **Scale**. As you make changes to these settings, the **Preview** area updates to give a visual preview of how the InDesign page will print on the selected paper size. Keeping an eye on the preview can save you from printing with inappropriate settings. If the page size of your document is larger than the paper size

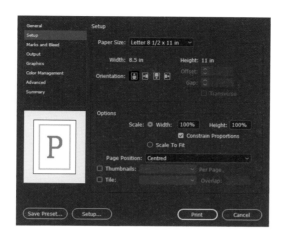

you want to print on, you can use the **Scale to Fit** option. InDesign scales the page, and any printer's marks and bleed, to fit onto the selected paper size.

Some print settings can be accessed through the printer's own dialog box as well as the InDesign Print dialog box. Where settings are duplicated, it is recommended that you use the InDesign settings rather than the printer's settings.

4 Marks and bleeds are used by commercial printers when preparing pages for the press, but you can also use the **Marks and Bleed** settings when printing composite proof copies of your pages – for example, you might sometimes print using Crop Marks so that you can trim a page to its final cut page size for

proofing purposes. Bear in mind that if you select any of the **Marks and Bleed** options, these add to the overall size of the printed area and, as a result, not all page marks and bleed objects may fit on your chosen page size.

If you are printing to a PostScript printing device, make sure you use the latest printer driver for your printer. Refer to the operating manual for complete information on setting up your printer correctly.

159

5 In the **Output** category, leave the **Color** pop-up set to **Composite RGB** to print a color composite proof. Select Composite Gray to

print a grayscale version of the document. If you are printing to

Controls for setting high-end print features such as color separations, Color Management and OPI (Open Prepress Interface) settings are not available for non-PostScript language printers.

...cont'd

If you import transparency effects into your document, or use the Effects panel, you need to consider which **Transparency Flattener** setting to use. To print transparent effects, InDesign divides overlapping areas into discrete segments, which are output as either vector or rasterized areas. The transparency flattener setting controls the balance between vector and bitmap (rasterized) information used to output these transparent areas.

Always advise your printer or service provider that you are using transparent effects in InDesign and ask for their recommendation as to which Flattener preset you should choose.

You can find further detailed information on transparency output issues on the **Flatten transparent artwork** pages of the InDesign User Guide website: **https://helpx.adobe.com/in/support/indesign.html**

a PostScript output device, you can choose **Separations** from the **Color** pop-up. You can then control advanced output settings such as trapping, half-tone screen settings and which inks print.

6 In the **Images** area of the **Graphics** category, the **Send Data** pop-up menu allows you to choose quality settings for images in your document. Leave this on **Optimized Subsampling** for a basic composite proof. For a PostScript printer, you may need to download fonts to the printer if you have used fonts in the document that are not resident on the printer itself. Leave the **Download** pop-up set to **Complete**.

7 In the **Advanced** category, choose a **Transparency Flattener** setting from the **Preset** pop-up if you have used Transparency or any effects such as Drop Shadow and Feathering from the Effects panel.

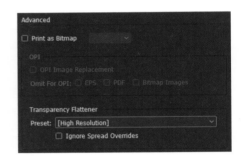

8 Click the **Setup** button to access the printer's own print settings dialog box. InDesign displays a warning box, which recommends that where possible you use settings from the InDesign **Print** dialog box, rather than from the printer's dialog box. Refer to your printer's instruction manual for information on the options available with your printer.

9 Click **Print** when you are satisfied with your settings.

Live Preflight

Preflighting a document, before handing it off to a service provider or printer, typically consists of checking for problems such as unwanted spot colors, images in the wrong color mode, overmatter, and so forth that might cause unwanted results or problems at output. The task is typically carried out as part of a final checking process, often by trained pre-press professionals.

InDesign performs a continuous "live" preflight as you create, edit and manipulate a document. You can set the conditions that constitute a "problem", and Live Preflight alerts you with a red circle in the Preflight readout at the bottom of the InDesign window when it detects problems.

The Preflight Panel
Use the Preflight panel to specify which Preflight Profile you want to use in a document and to find and fix problems identified by the profile.

1 To specify a Preflight Profile for the active document, use the Live Preflight pop-up menu to choose Preflight Panel, or choose Window > Output > Preflight (**Ctrl/Command + Alt + Shift + F**).

2 Make sure that the **On** checkbox is selected.

3 Use the **Profile** switcher pop-up menu to specify the preflight profile you want to use to check the document. As soon as you choose a new profile InDesign starts checking the document, indicated by the readout in the bottom-left corner of the dialog box. If there are errors in the document a red circle indicates this, along with a readout of the total number of errors. Errors are listed by category in the **Error** pane.

Creating Preflight Profiles
InDesign automatically applies the **[Basic]** preflight profile to new and converted documents.

Click on the **Live Preflight pop-up button**, along the bottom edge of the InDesign window, to reveal the Live Preflight pop-up menu:

Use the menu to switch Live Preflight On/Off for the active document only, or for all documents.

Live Preflight is switched on by default. The Live Preflight function can check for a wide range of conditions such as missing files or fonts, out-of-date files, images with the wrong color space and overset text, among others.

You can also use the Live Preflight pop-up menu to access the Preflight panel and the **Preflight Profiles** dialog box.

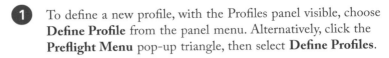

Use the **Pages** area in the bottom right of the Preflight panel to specify whether you want Live Preflight to check the entire document or a specific page range:

① To define a new profile, with the Profiles panel visible, choose **Define Profile** from the panel menu. Alternatively, click the **Preflight Menu** pop-up triangle, then select **Define Profiles**.

② Click the **New Preflight Profile** button (➕) on the left of the **Preflight Profiles** dialog box to begin creating a new profile. Enter a name for the new profile in the **Profile Name** entry box, replacing the highlighted placeholder name. The new name appears in the Profiles list on the left of the dialog box.

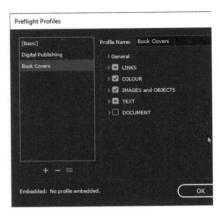

The **[Basic] profile** checks for missing or modified links, missing fonts and overset text frames. You cannot edit or delete this profile.

③ Use the **Expand** button (▶) to show options for each of the categories. There are six categories. Switch on the options you want to identify as problems, as required.

Examining and Fixing Problems

① Problems identified in the document are listed by category in the **Error** pane of the Preflight panel. Click the **Expand/Collapse** button to the left of a category or sub-category to reveal specific problems identified.

② Click on a problem in the list to select it. If necessary, click the **Info** expand button to display the **Info** area of the panel. An explanation of the problem for the selected entry appears in the Info area, with a suggested solution.

A dashed box (▣) to the left of a category name indicates that some settings are switched on for that category. A checkmark (☑) indicates that all settings within that category are switched **On**.

③ Click the hyperlink page number to the right of the entry in the Error pane to move to and highlight the specific problem in the document so that you can examine the issue and make decisions about how to fix it.

Package

The Package command is useful when you are preparing to send a document to your printer or service bureau. Packaging facilitates the process of bringing together the InDesign document and all image files and fonts used in the document – copying them into a folder, which you can then send.

1 Choose **File > Package**. You get a warning alert indicated by a warning triangle in the **Package** dialog box summary if InDesign detects any potential problems. You can either continue with

the packaging or cancel the **Package** dialog box and use the Preflight panel to locate and resolve issues with the document.

2 Select the **Create Printing Instructions** option in the **Package** dialog box if you want to supply contact details and any specific printing instructions to the printer. The printing instructions are saved into the package

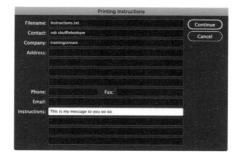

folder as a plain text file that can be opened by any text-editing application. The **Printing Instructions** panel appears after you click the **Package** button.

3 In the **Package Publication** dialog box, specify where you want to save the package folder, and specify the name for the package folder in the **Folder Name** entry box.

4 Select options to specify the set of items to be copied into the package folder.

When you click the **Package** button, you may be prompted to save the document before you can proceed to create the package:

To get more detailed information on aspects of the package, click the categories on the left of the **Package** dialog box:

163

To create a package for a Book publication you must choose **Package Book** from the Book panel menu (▤). If you have selected individual documents in the Book panel, you can choose **Package Selected Documents**.

...cont'd

Don't forget

The Package command generates a **Document fonts** folder.

InDesign can temporarily install fonts stored in a Document fonts folder at the same level as an open InDesign document. When the InDesign document is open, fonts from this folder take precedence over fonts of the same PostScript name installed elsewhere. When you close the InDesign document, fonts installed from the Document fonts folder are uninstalled.

Hot tip

It is only necessary to select the **Include Fonts and Links from Hidden Document Layers** option if there are additional layers in the document that contain these elements.

Hot tip

Select the **Include IDML** option if you want the Package command to generate a .idml version of the original InDesign document as part of the package. The IDML format allows for backward compatibility with older versions of InDesign.

Copy Fonts (Except CJK) – This copies the fonts required to print the document to a folder named "Document fonts" in the package folder.

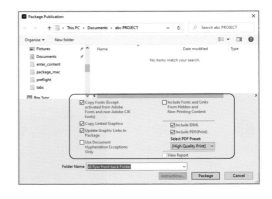

Copy Linked Graphics – This is an important option, as the original image file holds the complete file information needed for high-resolution printing. If you do not copy linked graphics to the folder, the document will print using low-resolution screen versions of images – typically producing poor-quality results. InDesign automatically creates a sub-folder named "Links" within the package folder when you select this option.

Update Graphic Links in Package – This allows InDesign to rewrite the paths of links to the images it copies to the Links folder.

Use Document Hyphenation Exceptions Only – This prevents the document from composing with the external user dictionary on the computer that opens the file. It is important to select this option when sending the document to a printer/service bureau.

Include PDF (Print) – This generates a PDF file of your publication as part of the package. Use the **Select PDF Preset** pop-up menu to select the preset you want to use. The presets available are the same as in the **Export Adobe PDF** dialog box when you export a PDF file (see pages 165-167).

 Click the **Package** button to start the process of copying files as necessary to the package folder. The **Font Alert** appears to remind you about copyright of fonts. Click **OK** if you are satisfied that you are not breaking any copyright agreements for the fonts in the document. InDesign gathers the files into the specified folder.

Exporting to PDF

You can export an InDesign document as PDF (Portable Document Format), either for high-resolution printing or for viewing using Acrobat Reader or web browsers. This section examines how you export documents for on-screen viewing.

One major advantage of using the PDF format is that it preserves the layout and content of the original InDesign document without the viewer needing to have access to InDesign itself. The other advantage is the small file sizes that PDF offers.

1 To export a file in PDF format, finalize your layouts and save the file. Choose **File** > **Export**. Choose **Adobe PDF (Print)** from the **Save as type** pop-up (Windows) or **Format** pop-up (Mac). Specify a location where you want to save the file, using standard Windows/Mac techniques. Enter a name for the file, and then click **Save**.

2 In the **Export PDF** dialog box, select an option from the **Adobe PDF Preset** pop-up menu. Each preset specifies a set of predefined settings, optimized for a particular PDF output requirement – for example, **[Press Quality]** is intended for PDF files that will be printed on imagesetters or platesetters as high-quality final output. **[Press Quality]** typically preserves the maximum amount of information contained in the original InDesign document. **[Smallest File Size]**, on the other hand, creates PDF files that are suitable for on-screen viewing – for example, on the World Wide Web.

[Smallest File Size] downsamples image quality and compresses

To avoid costly mistakes, make sure you preflight your document carefully and rigorously before you create a print-ready PDF file to send to your commercial printer.

To get the best results when you view a PDF file exported from InDesign, use **Acrobat Reader DC** or **Adobe Acrobat DC**.

To import a **PDF Preset** supplied to you by your printer or output bureau, choose **File** > **Adobe PDF Presets** > **Define**. In the **Adobe PDF Presets** dialog box, click the **Load** button, and then navigate to the **Presets** file. Click on the **Presets** file to select it, and then click **Open**:

...cont'd

If you are preparing PDFs for commercial printing, check with your printer or output bureau about which **Compatibility and Standard** settings to use.

Select the **Spreads** checkbox if you want left- and right-hand pages to be downloaded as a single spread. Use this option so that the PDF viewer displays spreads as if you are reading a magazine or book. **Do not** select the **Spreads** option for print publishing as this can prevent your commercial printer from imposing the pages.

After you choose a preset, if you select a category from the list on the left and then make changes to the predefined settings, "(modified)" is appended to the preset name.

file information to create a file that is as small as possible. When you choose one of the presets, settings in the PDF export categories change according to the preset you choose.

3 In the **General** category, specify whether you want to export all the pages in the document or a specified range of pages. In the **Options** area, select the **View PDF After Exporting** checkbox to launch the default PDF viewer – typically Acrobat Reader – so that you can check the result.

4 Compression settings are controlled initially by the preset you choose. Again, it is important if you are preparing a PDF file for commercial printing that you consult with your

printer or output bureau about which preset to choose, and only make changes to compression settings as directed.

5 You typically do not need to create **Marks and Bleed** settings for PDFs intended for on-screen viewing. For PDFs intended for print, ask your printer

or output bureau about the options you should set to meet their printing specifications.

6 In the **Output** category settings, if you are using the **[Smallest File Size]** preset, leave **Color** options set to the defaults.

Your commercial printer or output bureau may well have a PDF preset that they can supply that defines PDF export specifications to meet their specific requirements.

7 In the Advanced category, for printed final output and depending on the Compatibility setting

you are using, the **Transparency Flattener** setting is important; it can affect the quality of printed output if you have used effects such as Drop Shadow, Feathering or Opacity from the Effects panel. Always inform your printer the first time you use a new transparency effect, and check with them as to which Flattener setting you should choose.

If required, choose **Security** from the Categories list. Use the **Security settings** to control the degree of access that a viewer has for the PDF file – for example, you can set a Document Open Password so that only viewers who know the password are able to view the file. In the Permissions area, you can select the **Use a password...** checkbox to control editing and printing rights for the PDF.

8 Click **Export** when you are satisfied with the settings. If you chose the **View PDF after Export** option in the **General** category, Acrobat Reader launches and displays the exported PDF file.

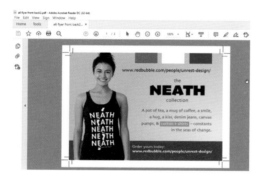

Saving a Preset
If you use a custom set of PDF settings on a regular basis, it is worth creating your own preset.

When you create PDF settings and then export the PDF file, the settings you created remain in force for other InDesign documents until you change them again.

1 To create a PDF preset, apply the settings you want in each of the PDF Export categories. Click the **Save Preset** button. Enter a name for the preset in the **Save Preset** dialog box, and click **OK**. The custom preset is now available in the **Presets** pop-up menu.

CC Libraries

CC (Creative Cloud) libraries provide a convenient and efficient way to store and manage page elements – such as frames (with or without content), groups, colors, and paragraph and character styles – that you need to reuse consistently for a publication, or across multiple publications.

You need an internet connection to create and manage libraries: CC libraries are cloud-based.

1. To create a new CC library, click the CC Library panel tab if it is in the active Panel Dock, or choose **Window > CC Libraries** to show the panel.

2. Click the **Add libraries** button at the bottom of the CC Libraries panel. Select **Create new library**. Enter a name for the library, then click the **Create** button to create a new empty library. You can now start to add elements to the library.

Click the **Back to libraries button** to move back from an active library to your list of libraries:

Adding Objects to a CC Library

1. To add an object to the active library panel, drag it into the middle of the library panel. Release when you see the **Add Artwork** cursor.

2. To control the specific attribute of a selected object that you place in the library, click the **Add Element** button at the bottom of the CC Libraries panel. Select an option from the pop-up menu. The list of options is dependent on the object you have selected.

A CC Libraries link icon (⌐△⌐) appears in the top-left corner of a selected frame (provided that Frame Edges are showing) and next to the object's entry in the Links panel (▰) to indicate that the object links to a CC library.

To add elements such as colors and paragraph and character styles to an InDesign document, right-click (Windows) or **Ctrl + Click** (Mac) on the element in the CC Libraries panel, then select the Add option to add the element to the appropriate panel.

Placing Objects from a CC Library

1. To place an object from a CC library onto an InDesign page, simply drag the element from the library onto the page.

2. Hold down **Alt/Option**, then drag a graphic/image from the CC library panel to embed the object in the InDesign document. The embedded object icon (▣) appears in the Layers panel to indicate an embedded object.

14 Transformations and Transparency

This chapter introduces the transformation tools. It also shows you how to change objects by adjusting transparency and by applying effects.

The Rotate Tool

The Rotate, Scale and Shear tools are located in the **Free Transform** Tool group:

You can also reposition the reference point marker by clicking on a **proxy reference point** in the Control or Transform panel:

In the Rotate dialog box, click the **Preview option** to see a preview of the transformation before you OK the dialog box.

The **Rotate tool**, like the **Scale** and **Shear** tools, works around a reference point – the point around which the transformation takes place. The reference point marker appears when you have a selected object and you then click on the **Rotate tool**. You can reposition the reference point marker if necessary.

1. To rotate an object, first select it with the **Selection tool**. Click on the **Rotate tool**. As soon as you select the **Rotate tool**, the reference point marker (⊹) appears on the object. The initial position of the marker is determined by the proxy reference point currently selected in the Control or Transform panels, or the Transform pane of the Properties panel. Position the Rotate cursor slightly away from the reference point marker, then drag in a circular direction. Hold down **Shift** as you drag to constrain the rotation to increments of 45°.

2. To reposition the reference point marker, with the **Rotate tool** selected, position your cursor on the marker, then drag it to another position. Alternatively, position your cursor at a different location, and then click. You can reposition the marker inside the selected object or anywhere on the page outside it.

3. To rotate using the Rotate dialog box, select an object using the **Selection tool**, then double-click the **Rotate tool**. Enter a rotation amount in the **Angle** entry field. Enter a value from -360 to 360. Negative values rotate an object in a clockwise direction; positive values act counterclockwise. This rotates the object and its contents.

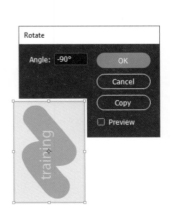

4 Click the **Copy** button instead of the **OK** button to create a rotated copy of the original object.

5 You can also use the **Rotate** field in the Control panel to rotate objects, or the contents of graphic frames.

6 To rotate the contents of a graphic frame, first select the content using the **Direct Selection tool**, then apply the rotation using the **Rotate tool**, the **Rotate** dialog box or the **Rotate** field in the Control panel.

7 You can use the **Rotate** dialog box to create a circular rotation effect. To create this effect, it is best to create a vertical and a horizontal ruler guide so that you can work easily around a center point. Start by creating a shape you want to rotate. Position it on the vertical guide above the center point. Make sure the shape remains selected. Select the **Rotate tool**. Position your cursor where the vertical and horizontal ruler guides meet. Hold down **Alt/Option**, and click. This does two things: it sets the reference point where you click, and it opens the **Rotate** dialog box. Enter a rotation angle, and then click the Copy button. Make sure the rotated object remains selected, then choose **Object** > **Transform Again** > **Transform Again** to repeat the transformation. Use the same command to continue repeating the transformation.

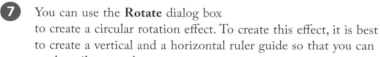

> **Hot tip**
> To quickly rotate a selected object, position your cursor just outside the corner of the object. When the cursor changes to the **Rotate cursor** (↱), drag in a circular direction.

> **Hot tip**
> When you are transforming objects or groups, the further away from the reference point marker you position the Transform cursor, the smoother the control you have over it.

> **Hot tip**
> For this technique to work effectively, you need to enter a rotation angle that divides into 360° – for example, 15, 30, 36 or 60:

The Scale Tool

You can scale objects manually using the **Scale tool**, or you can use the **Scale** dialog box. Like the **Rotate** and **Shear tools**, the **Scale tool** scales around a reference point.

When you use any of the transformation tools, if you press and drag the mouse in one action, you see a blue bounding box that represents the result of the transformation. When you release the mouse button, the transformation is applied to the object.

If you press the mouse button but pause briefly before you begin to drag, you see a live preview of the transformation on the object.

1 To scale an object manually using the **Scale tool**, first select the object with the **Selection tool**, then select the **Scale tool**. The reference point marker appears on the object or group. The position of the marker is determined by the currently selected proxy reference point in the Control or Transform panel, or the Transform pane of the Properties panel. (See page 34 for information on controlling the position of the reference point marker.)

2 Position the Scale cursor slightly away from the reference point marker, then start to drag. To scale the object in proportion, hold down **Shift** and start to drag at an angle of 45°. To scale the object horizontally only, hold down **Shift** and drag the cursor horizontally. To scale the object vertically only, hold down **Shift** and drag the cursor vertically.

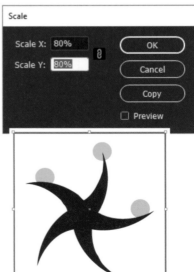

To scale the object but not its contents, use the Direct Selection tool to select all the Anchor points in the path, then apply the scaling:

3 To scale an object using the **Scale** dialog box, select the object using the **Selection tool**, then double-click the **Scale tool**. Make sure the **Constrain Proportions** button is selected, then enter a **Scale X** or **Y** amount to scale the object in proportion. Deselect the **Constrain Proportions** button and enter **Scale X** or **Y** values to scale non-proportionally.

The Shear Tool

Use the **Shear tool** to slant or shear an object. The **Shear tool** obeys the same basic principles as the **Rotate** and **Scale tools**.

The Shear tool is located in the Free Transform Tool group:

1 To shear an object manually, first select the object with the **Selection tool**, and then select the **Shear tool**. The reference point marker appears on the object or group. The position of the marker is determined by the currently selected proxy reference point in the Control or Transform panel. (See page 34 for information on controlling the position of the reference point marker.)

2 Position the **Shear cursor** slightly away from the reference point marker, and start to drag. Hold down the **Shift** key and drag at a 45° angle to constrain the shear. Hold down the **Alt/ Option** key as you shear to create a copy of the original object or group. You can use this technique to create shadow-like effects on objects.

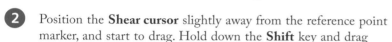

3 To use the **Shear** dialog box, select an object or group with the **Selection tool**, and then double-click the **Shear tool**. Enter a **Shear Angle** and select **Horizontal** or **Vertical** for the shear axis.

Shear

Shear Angle: -45

Axis
○ Horizontal
○ Vertical

OK
Cancel
Copy
☐ Preview

The Free Transform Tool

 When you rotate an object using the **Free Transform tool**, it rotates around its center point, regardless of which proxy reference point is selected in the Control or Transform panels, or the Transform pane in the Properties panel. (See page 34 for information on proxy reference points.)

 Using the **Free Transform tool**, press and drag in one move to see a wireframe representation of the effect of the transformation:

Press the mouse button down, pause briefly with the mouse button held down, then press and drag to see a full object preview of the effect of the transformation:

Unlike the Rotate, Scale, and Shear tools, the **Free Transform tool** does not display a reference point marker on the selected object when you select the tool. Using the **Free Transform tool** you can move, scale, rotate, reflect and shear objects. The tool's functionality is very similar to that of the equivalent tool in Photoshop and Illustrator.

1 To scale an object using the **Free Transform tool**, first select the object using the **Selection tool**, then select the **Free Transform tool**. Drag any selection handle to scale the object. Hold down **Shift** and drag a corner handle to scale the object in proportion. Hold down **Alt/Option** and drag a handle to scale the object around its center point.

2 To rotate a selected object, select the **Free Transform tool**, then position your cursor slightly outside the object's bounding box. The cursor changes to the Rotate cursor. Drag the cursor in a circular direction.

3 To reflect a selected object, select the **Free Transform tool**, then drag a handle through the opposite edge or handle.

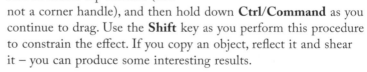

4 To shear a selected object, select the **Free Transform tool**, start to drag any of the center side or center top handles (but not a corner handle), and then hold down **Ctrl/Command** as you continue to drag. Use the **Shift** key as you perform this procedure to constrain the effect. If you copy an object, reflect it and shear it – you can produce some interesting results.

Transparency and Blending

Using transparency settings, you can allow underlying objects to show through other objects. You can apply transparency to selected objects, including graphic and text frames. You cannot apply transparency to individual text characters, but you can apply transparency selectively to an object's fill or stroke if required.

An opacity setting of 100% means that an object is completely solid. An opacity setting of 0% makes an object completely transparent.

① To set transparency for an object, select it using the **Selection tool**. Click the **Effects** icon if the panel is in the Panel Dock, or choose **Window > Effects (Ctrl/ Command + Shift + F10)** to show the Effects panel if it is not already showing.

② Either drag the Opacity slider or enter an Opacity amount. The lower the setting, the more transparent an object becomes.

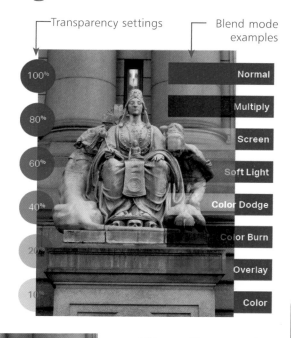

Transparency settings | Blend mode examples

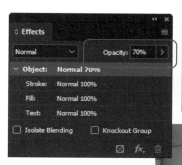

Hot tip

You can apply blend modes, such as Multiply and Color Burn, to objects using the **Blend Mode pop-up menu** in the Effects panel. When you apply a blend mode to an object that overlaps other objects, the color in the blend object mixes with the colors in the underlying object to produce a different color determined by the blend mode.

③ To overlay text on a faded or knocked-back area of an image, make sure you reduce the opacity for the fill of the text frame so that the opacity setting is not applied to the text.

Hot tip

To change Opacity or Blend mode selectively for either the stroke, fill or text of a selected object, make sure you select the appropriate Stroke, Fill or Text entry in the Effects panel before you change settings.

This is a caption placed on top of an image. The text frame has a reduced opacity setting.

When you use transparency settings, the Transparency Flattener setting you choose when printing or exporting PDFs can affect the final output.

A small transparency indicator ( ) appears alongside page thumbnails in the Pages panel to indicate that an object on the page has a transparency effect applied to it. You can control whether or not this icon appears by selecting **Panel Options** from the Pages panel menu (■).

You can use the **Flattener Preview panel** (Window > Output > Flattener Preview) to check for objects that use transparency.

You can also use the **Settings For pop-up menu** in the **Effects** dialog box to control where the effect is applied.

Effects

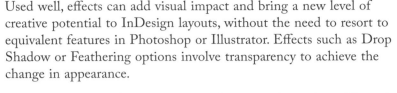

Used well, effects can add visual impact and bring a new level of creative potential to InDesign layouts, without the need to resort to equivalent features in Photoshop or Illustrator. Effects such as Drop Shadow or Feathering options involve transparency to achieve the change in appearance.

You can apply an effect to an entire object, or selectively to fill, stroke, or text in a text frame.

1 Click the **Effects panel icon**, or choose **Window > Effects** (**Ctrl/Command + Shift + F10**) to show the Effects panel.

2 Select an object, then select **Stroke**, **Fill** or **Text** if you want to apply the effect to a specific attribute of an object; otherwise, leave **Object** selected.

3 Click the **Add Effect** button (*fx.*) at the bottom of the Effects panel. You can also access the Effects menu by clicking the **Add Object Effect** button in the Control panel.

4 Click on the effect you want to apply to access the **Effects** dialog box.

...cont'd

5 The object effect you selected from the menu is selected by default, and settings available for the effect appear in the dialog box. Clicking in a checkbox switches the default effect **On** or **Off**. Click on the name of the effect itself to access the controls for that effect.

6 Select the **Preview** checkbox so that you can assess the changes you make to settings on the selected object. When you achieve the result you want, click **OK**.

7 Notice the 🔳 icon, which indicates that an effect is applied to an object, and whether it is applied to the object or a specific attribute of the object.

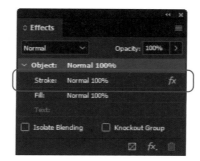

8 To edit settings for a selected object, double-click on the 🔳 icon in the Effects panel.

9 To remove all effects applied to a selected object and make it opaque (Opacity = 100%), click the **Clear Effects** button (▣). You can use commands in the Effects panel menu (▤) to clear effects and transparency selectively.

Consult with your commercial printer or service bureau when you start to use transparency settings: they can then make any recommendations so that you achieve satisfactory results. If possible, do a test run when you use these features for the first time.

You can apply more than one effect to an object. Click in a checkbox on the left of the **Effects** dialog box to add more effects. Click the name of the effect to adjust its settings.

177

Effects Gallery
There are nine transparency effects available in InDesign:

 No effect
 Drop Shadow
 Inner Shadow
 Outer Glow
 Inner Glow

 Bevel and Emboss
 Satin
 Basic Feather
 Directional Feather
 Gradient Feather

Gradient Feather Tool

The **Gradient Feather** tool provides a quick, convenient technique for applying and controlling one of the most popular InDesign effects – a gradient feather. Apply a gradient feather when you want to gradually fade an object to transparency – this can be a placed image, a group or any InDesign object.

To constrain the angle of a gradient feather to vertical, horizontal or increments of 45°, hold down **Shift**, then drag with the **Gradient Feather tool**.

1 Use the **Selection tool** to select an image, object or group. Select the **Gradient Feather** tool. Drag across the object to define the length of the fade to transparency and its angle.

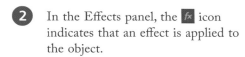

2 In the Effects panel, the *fx* icon indicates that an effect is applied to the object.

3 If you don't achieve the result you want the first time, you can simply drag with the **Gradient Feather tool** again to redefine the effect.

4 To edit the **Gradient Effect** settings using controls in the **Effects** dialog box, either double-click the *fx* icon or click the **fx** button (*fx*) at the bottom of the panel, then choose **Gradient Feather**.

Click the **Clear All Effects** button (▣) at the bottom of the Effects panel to remove all effects from the selected object.

5 Click the start (▫ – opaque) or end stop (▫ – transparent) on the transparency ramp, then drag to control the transition to transparency across the length of the effect defined with the **Gradient Feather tool**.

6 For a Linear gradient feather, use the Angle dial to control the angle of the feather. Select **Radial** from the **Type** pop-up menu to create a gradient feather that fades from the center out.

15 Paths and the Pen Tool

The Pen tool group, together with the Pathfinder commands, give you the potential to create and manipulate the paths of shapes and lines with exact precision and control.

Use the **Direct Selection tool** to select and edit points to define the exact shape of a path:

You create a closed path with the **Pen tool** when you click back at the start point.

When you are editing points and paths, if you find the **Anchored Object Control** disconcerting, you can hide it – choose **View** > **Extras** > **Hide Anchored Object Control**:

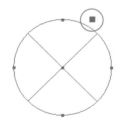

The word "point" is often used to mean **Anchor point**. **Direction points** are always referred to in full.

Points and Paths

In InDesign, the shape of all basic objects, including frames and shapes drawn with the **Pen** and **Pencil tools**, is defined by a path. You can manipulate paths and points in a variety of ways to achieve precisely the shape you need.

Paths and Points

A path consists of two or more **Anchor points** joined together by a curve or straight line segment. The **Pen tool** allows you to position **Anchor points** precisely where you want them as you create a path. You can also control which type of point you create – Smooth or Corner. The **Pencil tool** creates freeform paths, which are formed as you drag the cursor.

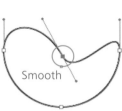

Open and Closed Paths

Using the **Pen** and **Pencil tools**, you can create open and closed paths. Objects such as rectangles and ovals, as well as text and graphic frames, are closed paths.

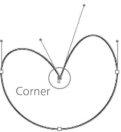

Closed path

Smooth and Corner Points

There are two kinds of points that you need to understand in order to work creatively and precisely with paths: **Smooth points** connect two curve segments in a smooth, continuous curve; **Corner points** allow an abrupt, sharp change in direction at the point. You can create paths consisting of both kinds of points as you draw them, and you can convert points from one type to the other using the **Convert Direction Point** tool.

Smooth

Corner

Direction Points

When you click on an **Anchor point** connecting curve segments, using the **Direct Selection tool**, one or two **Direction points** (depending on the type of point) appear, attached to their associated **Anchor point** by direction lines. **Direction points** control the length of a curve segment and the direction of the curve segment as it leaves the point.

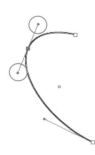

The Pen Tool

The **Pen tool** is the most versatile and precise tool you can use for defining shapes accurately. Use the **Pen tool** to create straight line segments, curve segments, or a mixture of both, with precise control over the positioning and type of **Anchor points**.

Straight Line Segments

① To create a straight line segment, select the **Pen tool**, position your cursor where you want the line to start, and then click. This sets the first **Anchor point**, and defines the start point of the path.

② Move the cursor to a new position. (Do not press and drag; simply reposition the cursor.) **Click**. This sets the second **Anchor point**. A straight line segment is created between the two points. Repeat the procedure to create the number of straight line segments you need.

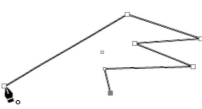

③ To finish drawing the path, position your cursor back at the start point (a small circle appears with the **Pen tool cursor**), and then click to create a closed path.

④ Alternatively, click the **Pen tool** again (or any other tool in the Toolbox) to create an open path. Clicking another tool indicates that the path is complete and that you have finished adding segments.

Curve Segments

① To create curve segments, with the **Pen tool** selected position your cursor where you want the path to start. Press and drag. This action sets the first **Anchor point** and defines its associated **Direction points**. (See pages 183–184 for further information on how **Direction points** control the shape of curves.)

Hold down **Shift**, and click with the **Pen tool** to constrain straight line segments to horizontal, vertical or increments of 45 degrees.

The **Center point** that appears when you select a path with the **Direct Selection tool** is not an editable point. It simply indicates the center of the bounding box that represents the overall dimensions of the path.

When you draw an open path with the **Pen** or **Pencil tool**, if there is a fill color selected, InDesign attempts to fill the path along an imaginary line from one end point to the other. This can be disconcerting at first. Click the **Fill** box and then click the **None** button to prevent this happening.

...cont'd

To make adjustments to a path while you are drawing it,

hold down **Ctrl/Command** to temporarily access the **Direct Selection tool**. Make changes as necessary; then release the **Ctrl/ Command** key to resume drawing the path.

To add more segments to an existing path, select an end point using the

Direct Selection tool. Select the **Pen tool**, then either click on the selected end point or press and drag on the end point. Move the cursor to a new position and continue to click, or press and drag, to add segments to the path.

You can also add and delete points on basic shapes such as rectangles, ovals and frames.

2 Release the mouse button. An **Anchor point** and two **Direction points** are visible.

3 Move the Pen tool cursor to a new position. Press and drag to set another **Anchor point** and to define its associated **Direction points**. Setting the second **Anchor point** also defines the curve segment between the first and second **Anchor points**. Repeat the procedure to create as many curve segments as you require.

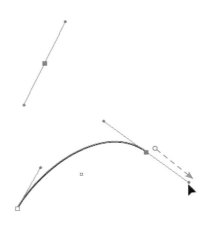

4 To finish drawing the path, position your cursor back at the start point (a small circle appears with the Pen tool cursor), then click to create a closed path. Alternatively, click the **Pen tool** again (or any other tool) to create an open path.

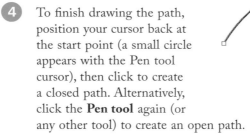

Adding and Deleting Points

Using the **Add** or **Delete Anchor Point tools**, or the **Pen tool**, you can add points to an existing path to achieve the exact shape you require, and you can delete points from a path to simplify it, if necessary.

1 To add an **Anchor point** to a selected path or frame, select the **Add Anchor Point tool**, or work with the **Pen tool**. Position your cursor on the path, then click to add a point. Points added to curve segments automatically appear with **Direction points**. Points added to straight line segments do not have **Direction points**.

2 To delete an existing **Anchor point**, select the path with either the **Selection tool** or the **Direct Selection tool**. Select the **Delete Anchor Point tool**, or work with the **Pen tool**. Position your cursor on a point, then click to delete the point. The path redraws without the point.

Selecting and Manipulating Points

To fine-tune the path you are working with, you need to select and manipulate **Anchor points** and **Direction points**. Use the **Direct Selection tool** to work on paths in this way.

Editing Anchor Points

1 Use the **Direct Selection tool**, then click on a path to select it. The path becomes highlighted and the **Anchor points** that form the shape appear as hollow squares. Click on an **Anchor point** to select it. The point becomes solid. If the **Anchor point** connects curve segments, **Direction points** also appear, connected to the **Anchor point** by direction lines. To change the shape of the path, drag the point to a new location.

When you have selected an **Anchor point** using the **Direct Selection tool**, you can press the Up/Down/Left/Right arrow keys on the keyboard to nudge the points in small steps.

X: 4in
Y: 3in

Editing Direction Points

Anchor points on curve segments have associated **Direction points** that control the length and shape of the curve segments. Continue working with the **Direct Selection tool** to edit **Direction points**.

Simply hovering over a path with the **Direct Selection tool** makes the path and its **Anchor points** available to edit without first clicking to select it.

1 To edit the **Direction points**, first select a point that has curve segments entering or leaving it. The **Anchor point** becomes solid and the associated **Direction points** appear, connected to the **Anchor point** by direction lines.

2 Position your cursor on a **Direction point**. Drag the **Direction point** further away from the **Anchor point** to increase the length of the curve segment it controls. Drag the **Direction point** closer to the **Anchor point** to make the curve segment shorter.

Hold down **Shift** as you drag **Anchor** and **Direction points** to constrain the movement to vertical, horizontal or multiples of 45°.

3 Drag a **Direction point** in a circular direction around the **Anchor point** to change the angle at which the curve segment enters or leaves the **Anchor point**. The result is to change the shape of the curve segment.

Smooth and Corner Points

Understanding the difference between **Smooth** and **Corner Anchor points** will give you complete control over the shape of paths.

Smooth Points

A **Smooth point** maintains a smooth, continuous transition or curve from the incoming to the outgoing curve segments, through the **Anchor point**.

1 Select the **Direct Selection tool**, then click on a **Smooth Anchor point**. Two **Direction points** appear, connected to the **Anchor point** by direction lines. Position your cursor on a **Direction point**, then press and drag in a circular direction around the point. As you move the **Direction point**, the opposite **Direction point** moves like a balance, keeping both **Direction points** perfectly aligned. If you move the **Direction point** further away from or closer to the **Anchor point**, the distance of the opposing **Direction point** does not change.

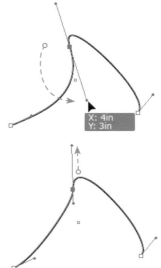

Corner Points

Use **Corner points** to create a sharp change in direction at the **Anchor point**.

1 Click on a **Corner point** to select it. Two **Direction points** appear, connected to the **Anchor point** by direction lines. Drag a **Direction point** in a circular direction around the **Anchor point** and/or move the **Direction point** closer to or further away from the **Anchor point**. The opposite **Direction point** does not move: for **Corner points**, each **Direction point** works completely independently of the other.

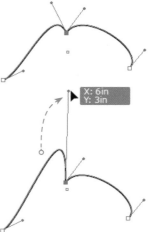

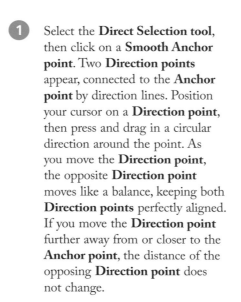

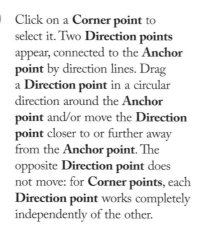

Converting Points and Cutting Paths

The **Convert Direction Point tool** can convert a **Smooth point** to a **Corner point** and vice versa, and it can also be used to retract **Direction points** or create **Smooth points**.

Converting Smooth to Corner

1 To convert a **Smooth point** to a **Corner point**, select the point using the **Direct Selection tool**. Select the **Convert Direction Point tool**. Position your cursor on one of the **Direction points**, then press and drag. The point becomes a **Corner point** – each **Direction point** moves independently of the other.

2 If you are going to make further changes to the **Direction points** to fine-tune the curves, make sure you reselect the **Direct Selection tool**, rather than continuing to work with the **Convert tool**.

Converting Corner to Smooth

1 To convert a **Corner point** to a **Smooth point**, select a **Corner point** using the **Direct Selection tool**. Select the **Convert Direction Point tool**. Position the cursor on the **Anchor point** (not a **Direction point**), then drag off the point to convert the point to a **Smooth point** and to define the shape of the smooth curve.

2 Reselect the **Direct Selection tool** to make further changes to the point or **Direction point**. If you continue to work with the **Convert tool** on the same point, or its **Direction points,** you will undo the results of Step 1.

While working with the Direct Selection tool, you can temporarily access the Convert Direction Point tool by holding down Ctrl/ Command + Alt/Option.

You can use the Convert Direction Point techniques on the Anchor points that define the shape of a text or graphic frame:

The bottom row of the Pathfinder panel includes buttons for converting points. Choose Window > Object & Layout > Pathfinder to display the panel.

...cont'd

Using the **Convert Direction Point tool**, you can click on a **Direction point** to retract that individual point.

If you split a picture frame containing an image, you end up with a copy of the image inside each half of the frame.

After you cut a path, select the **Direct Selection tool** to make further changes to either side of the cut path if required.

Retracting Direction Points

1. To retract both **Direction points** for either a **Smooth point** or a **Corner point**, first select the point with the **Direct Selection tool** then select the **Convert Direction Point tool**. Position your cursor on the **Anchor point**, then click to retract the **Direction points**. The incoming and outgoing curve segments are redrawn accordingly.

Converting a Retracted Point to a Smooth Point

1. To convert a **Retracted point** to a **Smooth point**, select a path with the **Direct Selection tool**. A **Retracted point** is one that has no **Direction points** associated with it when you click on it with the **Direct Selection tool**. Select the **Convert Direction Point tool**. Position your cursor on the **Retracted point**, then drag off the point to create a **Smooth point** and define the shape of the incoming and outgoing curve segments.

The Scissors Tool

Use the **Scissors tool** to split or cut a path anywhere along a curve or straight line segment, or at an **Anchor point**.

1. To split an open or closed path, select the **Scissors tool**. Position your cursor at the point on the path where you want to cut it. (The path does not have to be selected, and you do not have to click on an existing **Anchor point**.) **Click**. Two **Anchor points** are created at the point at which you click.

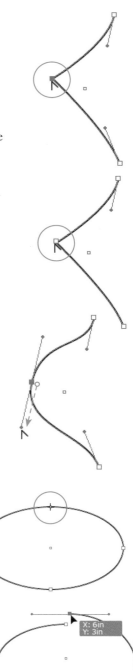

Pathfinder Commands

Use the Pathfinder commands to create new shapes from overlapping frames or shapes. The resultant paths can be interesting, complex shapes that would be difficult to create any other way.

Choose **Window > Object & Layout > Pathfinder** to show the Pathfinder panel if it is not already showing.

Add

The **Add pathfinder** command creates a more complex shape from overlapping shapes. **Add** is useful when you want to create a complex shape with a unified outline or stroke from two or more basic shapes.

 To add shapes, make sure you select at least two frames or basic shapes. Click the **Add** button

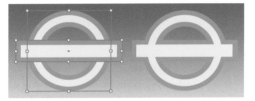

on the Pathfinder panel, or choose **Object > Pathfinder > Add**. When frames or shapes have different fill and stroke attributes, the Add command applies the fill and stroke attributes of the frontmost object to the resultant shape.

Typically, the Pathfinder commands work by creating new shapes where existing paths overlap.

Subtract

The **Subtract** command acts like a punch – shapes in front of the backmost object punch through and cut away areas of the backmost object where they overlap. The frontmost objects are deleted when you use the command. This is a useful technique for creating completely transparent areas in a shape.

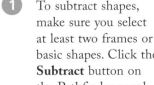 To subtract shapes, make sure you select at least two frames or basic shapes. Click the **Subtract** button on the Pathfinder panel, or choose **Object >**

Pathfinder > Subtract. When frames or shapes have different fill and stroke attributes, the backmost shape retains its fill and stroke attributes when you use the **Subtract** command.

You must have **two** overlapping shapes or frames selected to use commands in the Pathfinder panel.

...cont'd

To quickly transform one shape to another shape, click one of the **Convert Shape buttons** in the Pathfinder panel.

Intersect

The **Intersect** command creates a new shape where two shapes or frames overlap. Areas that do not overlap are removed. You can use the command for only two objects at a time. If you attempt to apply the command with more than two objects selected, a warning prompt indicates that you cannot proceed.

1 To intersect shapes, select two frames or basic shapes, click the **Intersect** button in the Pathfinder panel, or choose **Object > Pathfinder > Intersect**. The resultant shape retains the fill and stroke attributes of the frontmost object.

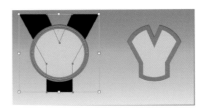

Exclude Overlap

The **Exclude Overlap** command makes the area where two or more frames or shapes overlap completely transparent.

To use the Pathfinder commands on text, you must first convert it to outlines. Select a text frame using the **Selection tool**, then choose **Type > Create Outlines**:

1 To use **Exclude Overlap**, select two or more frames or basic shapes, then click the **Exclude Overlap** button in the Pathfinder panel, or choose **Object > Pathfinder > Exclude Overlap**. The resultant shape retains the fill and stroke attributes of the frontmost object.

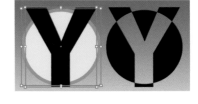

Minus Back

Minus Back is the opposite of the **Subtract** command. Objects behind cut away the frontmost object where they overlap.

1 To use Minus Back, select two or more frames or basic shapes, then click the **Minus Back** button in the Pathfinder panel, or choose **Object > Pathfinder > Minus Back**. The resultant shape retains the fill and stroke attributes of the frontmost object.

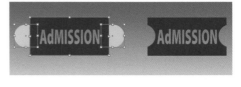

Index

U

V

W

X

Z